Hot Science is a series exploring the cutting edge of science and technology. With topics from big data to rewilding, dark matter to gene editing, these are books for popular science readers who like to go that little bit deeper ...

AVAILABLE NOW AND COMING SOON:

Destination Mars:
The Story of Our Quest to Conquer the Red Planet

Big Data:
How the Information Revolution
is Transforming Our Lives

Gravitational Waves:
How Einstein's Spacetime Ripples Reveal the Secrets
of the Universe

The Graphene Revolution:
The Weird Science of the Ultrathin

CERN and the Higgs Boson:
The Global Quest for the Building Blocks of Reality

Cosmic Impact:
Understanding the Threat to Earth from Asteroids
and Comets

Artificial Intelligence:
Modern Magic or Dangerous Future?

The Searc   erse

Dark Matter & Dark Energy:
The Hidden 95% of the Universe

Outbreaks & Epidemics:
Battling Infection From Measles to Coronavirus

Rewilding:
The Radical New Science of Ecological Recovery

Hacking the Code of Life:
How Gene Editing Will Rewrite Our Futures

Origins of the Universe:
The Cosmic Microwave Background
and the Search for Quantum Gravity

Behavioural Economics:
Psychology, Neuroscience,
and the Human Side of Economics

Quantum Computing:
The Transformative Technology of the Qubit Revolution

The Space Business:
From Hotels in Orbit to Mining the Moon
– How Private Enterprise is Transforming Space

Game Theory:
Understanding the Mathematics of Life

Hot Science series editor: Brian Clegg

THE SPACE
BUSINESS

ABOUT THE AUTHOR

Andrew May is a freelance writer and science consultant. *The Space Business* is his fourth book in the Hot Science series, following *Destination Mars, Cosmic Impact,* and *Astrobiology*. He lives in Somerset.

THE SPACE BUSINESS

From Hotels in Orbit to Mining the Moon – How Private Enterprise is Transforming Space

ANDREW MAY

ICON

Published in the UK and USA in 2021
by Icon Books Ltd, Omnibus Business Centre,
39–41 North Road, London N7 9DP
email: info@iconbooks.com
www.iconbooks.com

Sold in the UK, Europe and Asia
by Faber & Faber Ltd, Bloomsbury House,
74–77 Great Russell Street,
London WC1B 3DA or their agents

Distributed in the UK, Europe and Asia
by Grantham Book Services,
Trent Road, Grantham NG31 7XQ

Distributed in the USA
by Publishers Group West,
1700 Fourth Street, Berkeley, CA 94710

Distributed in Australia and New Zealand
by Allen & Unwin Pty Ltd,
PO Box 8500, 83 Alexander Street,
Crows Nest, NSW 2065

Distributed in South Africa
by Jonathan Ball, Office B4, The District,
41 Sir Lowry Road, Woodstock 7925

Distributed in India by Penguin Books India,
7th Floor, Infinity Tower – C, DLF Cyber City,
Gurgaon 122002, Haryana

Distributed in Canada by Publishers Group Canada,
76 Stafford Street, Unit 300
Toronto, Ontario M6J 2S1

ISBN: 978-178578-745-4

Typeset in Iowan by Marie Doherty

Printed and bound in Great Britain
by Clays Ltd, Elcograf S.p.A.

CONTENTS

SPACE FOR EVERYONE? 1

Imagine saving up for the trip of a lifetime to Sky Hotel, orbiting 16,000 kilometres above the surface of the Earth. There, you can indulge in a choice of activities, from familiar ones like cordon bleu dining or gambling in the casino, to the very far from usual, such as zero-gravity sports. Or how about swimming round and round on the inside of a rotating cylinder, where the water is held in place by centrifugal force? Then there are the views of Earth – which you can see in its stunning entirety, from pole to pole, or magnified through a telescope so you can see individual buildings in any city you choose to focus on.

That's the scenario that Arthur C. Clarke described in his article 'Vacation in Vacuum', published in *Holiday* magazine way back in 1953. Best known as a science fiction author, Clarke was also a scientist and visionary who was among the first to grasp the real-world possibilities of space travel. In 1945 he famously championed the idea of geosynchronous communication satellites, two decades before they became a reality. His key realisation was that space isn't just interesting, it's *useful*. When the space race turned its attention to the Moon in the 1960s, most people saw it as a purely political goal – or at best an exercise in pure science. Clarke was one of the few who could see further than that. In his 1966 book *Voices from the Sky*, he wrote of the Moon that:

> A century from now it may be an asset more valuable than the wheatfields of Kansas or the oil wells of Oklahoma – an asset in terms of actual hard cash, not the vast imponderables of adventure, romance, artistic inspiration and scientific knowledge.

The idea of making money *from* travelling to the Moon – rather than sinking billions of dollars into simply planting a flag on its surface – was a novelty, and one that few people outside the world of science fiction took seriously. Yet the basic concept was as sound then as it is now. Take lunar tourism, for example – the basis of Clarke's novel *A Fall of Moondust* (1961). If people are going to splurge out for a vacation in Earth orbit, they'll splurge even more for one on what is effectively a whole different world.

There are other types of space business that don't depend on milking super-rich customers. They just make hard-nosed economic sense, or they will do once they're established. That's true, for example, of another of the topics discussed by Clarke, lunar mining. As we'll see later, there are valuable elements that are far easier to extract on the Moon than here on Earth. The same is true – maybe even more so – of near-Earth asteroids. A staple of science fiction since the early 20th century, asteroid mining has the potential to be one of the most lucrative undertakings beyond the Earth's atmosphere.

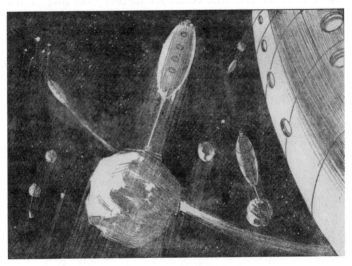

A vision of asteroid mining from the December 1935 issue of *Amazing Stories*.

(Public domain image)

Sixty years after the first humans were launched into orbit, we still haven't quite achieved Arthur C. Clarke's vision of a thriving, self-sustaining space business – but we're getting there. Although communication satellites have been around since the 1960s, it's only in recent years that they've been built, launched and operated entirely by private companies – and in a big way, too. Just think of Elon Musk's

SpaceX and its vast constellation of Starlink satellites, which aim to bring broadband internet to remote locations all over the world.

Similarly, we're now seeing the birth of the first privately operated space tourism companies, such as Richard Branson's Virgin Galactic and Jeff Bezos's Blue Origin, with their modest offering of brief suborbital hops. Far more ambitious plans – including orbiting hotels and trips around the Moon – are in advanced stages of preparation. Other companies are working on the technology needed to extract minerals from asteroids, or to operate robotic mining equipment on the Moon.

All these topics – space tourism, private satellite constellations, asteroid mining and more – will be discussed in detail in the chapters to come. First, however, we need to address one very obvious question. Why is the space business taking so long to get up and running?

Space Is Hard

You may have heard the phrase 'space is hard', because it's become something of a cliché. Richard Branson said it in the wake of the fatal crash of Virgin Galactic's SpaceShipTwo spaceplane during a test flight in October 2014. The following June, the phrase was used by astronaut Scott Kelly on board the International Space Station (ISS), after a SpaceX cargo craft was destroyed en route to the station. And Peter Diamandis, the founder of the Lunar X Prize for the first private company to put a robotic lander on the Moon, said the same thing when the most promising contender, Israel's Beresheet, crashed onto the lunar surface in April 2019.

The fact that the phrase gets used so often, and under circumstances like these, more or less proves that it's true. Space really is hard, for a variety of reasons. It involves highly complex – and often new and untested – technology, hence the frequent mishaps and accidents. It's an immensely expensive business, particularly in the developmental stages, so even wealthy companies can struggle to get the necessary funding together. Hardest of all, it involves doing things that evolution just hasn't prepared humans for, such as ascending for hundreds of kilometres against the pull of Earth's gravity, or surviving in the vacuum of outer space.

Up to a certain point, there's no fundamental physics preventing us reaching higher and higher altitudes. Both helium balloons and jet aircraft, if they're specially designed for the task, can climb to 30 kilometres or a little beyond that. But that's when the problems start, because the higher you get, the less atmosphere there is to support you. At 40 km the air density is only a 300th of its value at sea level, and at twice that height it is 200 times smaller still.

It's easy to see why a helium balloon has an altitude limit. The balloon starts to rise because, at sea level, it's lighter than the air it displaces. The mass of the gas inside the balloon is less than the mass of the same volume of out-side air. But as it rises and the air density decreases, there comes a point when that's no longer true – and the balloon stops rising.

The situation with a jet plane is a little more complicated, because it involves two different effects. Unlike a balloon, a fixed-wing aircraft is heavier than air, but it's still able to rise due to the aerodynamic lift produced by the flow of air over its wings. But for that to work, there has to be enough air in the first place. So producing lift becomes harder and harder as the surrounding atmosphere gets thinner.

A jet needs the atmosphere for another reason, too. Its forward thrust is produced by pulling large quantities of air into its engines, and using the oxygen in it to burn fuel and drive a turbine – which then blasts out a fast-moving stream of exhaust which pushes the jet along. This too ceases to work at high altitudes, where there simply isn't enough atmosphere.

By convention,* space starts at an altitude of 100 km, known as the 'Kármán line'. That's a nice round figure, and with an air density more than 2 million times smaller than at sea level, few people would dispute that for all practical purposes it's outside the atmosphere. But there's a more concrete reason why the pioneering aeronautical engineer Theodore von Kármán picked that particular value. He calculated that for an aircraft to stay aloft at that altitude through aerodynamic lift, it would have to travel at orbital velocity (a concept we'll explore in more detail shortly) – and it would then stay aloft anyway, even in a total vacuum.

If balloons and jet aircraft are out, then, the only realistic way to get to the Kármán line – and beyond – is with the aid of a rocket. This works on the same physical principle as a

---

* By international convention, that is. The US government uses an older definition by which space starts at 50 miles (80 km) altitude.

jet, blasting out a fast-moving stream of gas in the opposite direction to the one you want to travel in. The difference is that a rocket is entirely self-contained. While a jet can get most of the working material it needs from the surrounding atmosphere, mixing it with a relatively small proportion of fuel to give it the necessary energy, a rocket typically has to carry all its fuel and propellant along with it.

Actually, once you've cracked the problem of building a working rocket, simply getting into space – beyond the Kármán line – isn't really that hard at all. The difficult part is staying up there without falling straight back to Earth. To see this, you only have to consider the curious case of MW 18014 – the curious thing about which is that it's hardly ever mentioned in the history books.

MW 18014 was a V-2 – which, in a more generic sense, certainly hasn't been forgotten by history. The first rocket powerful enough to carry a substantial payload over a distance of hundreds of kilometres, the V-2 wasn't designed as a space launcher but as a weapon of war. It entered service with the German army in 1944, and was used, among other things, to attack London from launch sites on the Dutch coast – a distance of some 300 km.

Unlike an aircraft, the V-2 wasn't powered throughout its flight – only during a short initial boost phase to get it up to the desired speed – after which its own momentum kept it going. Its trajectory was similar to the parabola that a ball follows when you throw it. If you throw the ball up at a steep angle, it goes very high, but doesn't travel very far horizontally before falling back to the ground. Conversely, if you throw it at a shallow angle, the horizontal range will be much longer, but the maximum height reached will be lower.

The V-2's designers were faced with a similar trade-off. Because the rocket didn't rely on aerodynamics to keep it aloft, it made sense to fly as much of the trajectory as possible in the thin upper atmosphere, in order to minimise drag forces. To achieve the desired range with a full fuel load, the highest point of the parabola – technically called the apogee – worked out at around 80 kilometres, much higher than any aircraft or balloon had flown.

In the case of a ball, you know intuitively that the way to get it as high as possible is to throw it vertically upwards. It's the same with rockets, and that's where MW 18014 comes in. It was a test launch that took place on 20 June 1944, several months before the V-2 entered military service, at the army's research establishment at Peenemünde on Germany's Baltic coast. Unlike an operational mission, this particular rocket was fired vertically upwards, in order to check that it still functioned correctly at very high altitudes. It hit apogee at 176 km, well above the Kármán line – a feat that meant it was the very first human-made object to reach outer space.

Yet MW 18014 didn't make the history books. No one claims the space age started on 20 June 1944. The reason is that, having reached that height of 176 km, MW 18014 immediately started falling, to plunge ignominiously into the depths of the Baltic. When it reached apogee its speed was zero, after which it was entirely at the mercy of gravity. And gravity, in its inimitable way, pulled MW 18014 back down to Earth.

Things would have been different if the rocket had been launched at an angle rather than vertically. In this case it would have a horizontal component of velocity as well as a vertical one, and even at apogee it would still be moving parallel to the surface of the Earth. Of course, if exactly the same rocket was used it wouldn't be able to reach the same record-breaking altitude on a non-vertical trajectory. For the sake of argument, then, let's assume the rocket has a more powerful engine, allowing it to reach the same altitude regardless of its horizontal motion.

Gravity will still pull the rocket back down, but the rocket will travel horizontally in the process, eventually crashing at some distance from the launch site. Given enough horizontal speed, it might even have reached London, which is much further from Peenemünde than from the operational launch sites – about a thousand kilometres, in fact. That's far enough, in relation to the radius of the Earth, that the flight path would actually curve around the surface of the planet. If the rocket was even more powerful, allowing it to reach an even greater horizontal speed, it would curve that much further round the Earth before coming back down.

You've probably noticed that if you drive too fast round a corner, you feel a force pushing you in the opposite direction to the turn. It's called centrifugal force, and some people will tell you it isn't a real force at all, but an illusory one produced by your own inertia. From your point of view inside the car, however, it's real enough – and, in the same way, it's real enough to a rocket as it flies around the curvature of the Earth. In the latter case, if the rocket travels fast enough

– around 28,000 kilometres per hour if it's just above the Kármán line – the centrifugal force is strong enough that it exactly counteracts the downward pull of gravity. The result, which sounds like black magic no matter how many times you hear it, is that the rocket *never* falls back to Earth. It's in orbit.

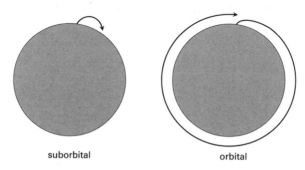

suborbital

orbital

The difference between a suborbital rocket launch and an orbital one is the much higher velocity required by the latter.

It's at this point that 'space is easy' suddenly becomes 'space is hard', because 28,000 kph is an eye-wateringly high speed. You may remember from high school physics that potential energy – the energy required to raise an object through height h – is proportional to h, while kinetic energy – the energy to increase its speed by v – is proportional to v *squared*. In the present context, this means the energy needed to get a payload up to the Kármán line, 100 km above the Earth's surface, is nothing compared to the energy needed to get the same payload up to orbital velocity.

It will be easier to visualise the difference we're talking

about if we return to the ball-throwing analogy. Rather than throwing it all the way up to the Kármán line, let's scale things down by a factor of 10,000. That brings the altitude down to ten metres, which a reasonably fit person might manage quite easily. Working out the required speed is a little harder, because that factor of 10,000 applies to v^2 rather than v, so we only have to divide orbital velocity by the square root of 10,000, which is 100. The bottom line is that, if reaching the Kármán line is like throwing a ball 10 metres up in the air, then reaching orbital speed would involve throwing it with the superhumanly fast speed of 280 kph. That requires far more energy, by a factor of 30 or so, than the first throw.

Having to accelerate a rocket all the way to orbital speed introduces new problems that the V-2 designers never had to contend with. More acceleration means more energy, and more energy means more fuel. But more fuel means more mass, and more mass means it needs *even more energy* to get to the desired speed. It's a vicious cycle that gives rocket scientists nightmares, and even today, no single-stage rocket has ever made it into orbit. Instead, they have to be constructed in multiple stages – two or more – with the earlier stages dropping off as their fuel is used up in order to minimise the mass that has to be shifted.

After MW 18014's success in breaching the Kármán line, it took another thirteen years of intensive effort by teams on both sides of the planet – the United States and what was then the Soviet Union – before an object was put into space that actually stayed up there. This was the Soviet satellite

Sputnik 1, launched on 4 October 1957 – and it was the real start of the space age.

An Exclusive Club

Sputnik 1 had two objectives, one stated and one unstated. The stated objective was scientific research, the unstated one was a show of military strength. The rocket that launched the satellite into orbit was a repurposed intercontinental ballistic missile, a Cold War status symbol that only the Soviet Union possessed in 1957. As different as they are, these two objectives have at least one thing in common – they're both essentially preoccupations of government, not the private sector. Doing pure science, like building up weapon stockpiles, is something that costs money – it doesn't make money.

This set the tone for the space sector for decades to come. It was a government monopoly – initially just the great Cold War rivals of the Soviet Union and United States, then with other countries following suit. By the 1990s, Japan, China, India, Israel and the European Union also had space launch capabilities – and in every case, it was government-owned, government-funded and government-operated.

This isn't a bad thing per se, because governments are able to pump money into developing highly complex technology where the lead times are measured in years or decades, which would quickly bankrupt any private company. But it did leave the world with the impression that space was only useful for doing government-like things, and that it was destined always to remain the exclusive preserve of governments.

Of course, those governments accomplished some remarkable feats in the last four decades of the 20th century – not least the advent of human spaceflight. Three and a half years after Sputnik 1, on 12 April 1961, an upgraded version of the same launch rocket was used to send the first person into orbit, Major Yuri Gagarin of the Soviet Air Force. As with Sputnik, this was seen around the world as another high-profile victory for the Soviet side in the Cold War.

The US response to Gagarin's flight was interesting. At the time, they didn't have the capability to launch an American into orbit – that had to wait until John Glenn's flight in February 1962. But they reasoned that as far as the history books were concerned, 'space' meant anywhere beyond the Kármán line, not necessarily an orbital mission. So they did the best they could, with what was essentially a piloted version of MW 18014.

Alan Shepard, the first American in space, was launched atop a single-engined, single-stage Redstone rocket – only slightly more powerful than a V-2 – from Cape Canaveral in Florida at 9.34 a.m. local time on 5 May 1961. Around two and a half minutes later, when the Redstone had completed the boost phase, Shepard's three-metre-long Mercury capsule separated and continued upward on a parabolic trajectory. It reached apogee at 188 km – only a dozen kilometres higher than MW 18014 – and then gravity took over and it started to fall back to Earth. Slowed by parachutes, the capsule splashed down in the Atlantic about 400 km off the Florida coast. The time was 9.49 a.m.; the whole flight had taken just fifteen minutes.

It's tempting to be cynical about Alan Shepard's place in the history books, given that MW 18014 – which did much the same thing seventeen years earlier – is almost completely forgotten. That's not really fair, though, because it ignores the importance of the human angle. The V-2 engineers merely witnessed MW 18014's flight into space, whereas Shepard experienced it for himself. He saw the Earth from much the same altitude that Yuri Gagarin did, albeit for a much shorter time. And, unlike MW 18014, he survived the flight. He travelled in the same type of Mercury capsule John Glenn would use on his orbital mission, which protected him from the vacuum of space and brought him safely back to Earth.

The fact is that Alan Shepard really did travel into space, and with significantly less sophisticated technology than that required for an orbital flight. If you're a thrill seeker in search of a genuine space experience, and you want to keep your costs to a bare minimum, there's only one way to do it – by following in the footsteps of Alan Shepard. There's a genuine market here, because the world has never had a shortage of wealthy thrill seekers. So it's astonishing, in hindsight, that it took entrepreneurs so long to spot a golden opportunity. It's only in recent years that they've finally cottoned on, and people like Jeff Bezos and Richard Branson are planning to offer Shepard-style 'suborbital' flights to budding space tourists – as we'll see in the next chapter.

On the other hand, back in the 1960s, it doesn't seem to have crossed anyone's mind that space travel might be of serious interest to anyone other than powerful nation states. More narrowly still, it was often seen exclusively in the context of the bitter Cold War rivalry between the USA and the USSR. This is demonstrated as clearly as anywhere in the first piece of international space legislation, the Outer Space Treaty of 1967. Its formal title is 'Treaty on Principles Governing the Activities of States in the Exploration and Use of Outer Space, including the Moon and Other Celestial Bodies' – and that use of the word 'states', rather than a more general term, is telling. The treaty was put together against the backdrop of Cold War paranoia, and its main focus was on limiting the military use of space.

That's an admirable enough goal, and the treaty did have the laudable effect of preventing nuclear weapons being deployed in space. But in other ways it was short-sighted to the point of incompetence. By repeatedly emphasising that word 'state', it gave the impression that national space programmes, rather than private ones, would always be the norm. At one point it even says that 'states shall be responsible for all national space activities whether carried out by governmental or non-governmental entities'. It then says, 'outer space is not subject to national appropriation by claim of sovereignty, by means of use or exploitation, or by any other means.'

The Outer Space Treaty was an idealistic exercise in world harmony, at a time when the animosity between East and West was as intense as it has ever been. When Apollo 11 landed on the Moon in 1969, the astronauts famously planted the Stars and Stripes – not to claim sovereignty, but to symbolise America's victory in this particular race against the Soviets. Immediately before raising the flag, they unveiled a plaque that read 'We Came in Peace for all Mankind'. This wasn't just a nice gesture, it was required by a clause of the Outer Space Treaty: 'The exploration and use of outer space shall be carried out for the benefit and interests of all countries and shall be the province of all mankind.'

In those days, the 'benefit and interests' of space exploration were seen almost exclusively in scientific rather than commercial terms. It's true that valuable science can be done in space, and in fact a lot of great science really was done by the Apollo missions. Contrary to popular belief, they weren't all about making footprints and planting flags. By bringing back over a third of a tonne of rock samples from various parts of the lunar surface, and by installing a network of seismometers and other instruments, the astronauts added an enormous amount to our understanding of the Moon, and of the Solar System in general.

The problem is, even in terms of its scientific goals – let alone its political ones – the Apollo programme was never anything but a short-term project. It had a number of specific objectives, and when these had been met the whole thing came to a dead end. NASA went on to pursue other goals in space, such as the Shuttle and the International Space Station (ISS), with little serious thought for anything else the Moon might be used for. Occasionally there was talk of establishing a more permanent lunar base, but this would have been a scientific research station, funded entirely by the taxpayer – not a mining operation or tourist resort that might ultimately have paid its own way.

There's a fundamental difference between the economics of the public and private sectors. Private companies only use money with the aim of ultimately making more money. They have to, otherwise they'd go bust. The government, on the other hand, uses money in whatever ways will go down well with the voting public. When it came to sending humans to the Moon, there was just that one narrow window in the 1960s when the public was behind the idea. Both the Americans and Russians were desperate for a decisive victory over the other side, but it had to be done in a way that didn't involve nuclear armageddon. A race to the Moon was the obvious answer, and never mind the cost.

Treating Apollo as a high-stakes race against time, in which cost-effectiveness was a secondary priority at best, led inevitably to an eye-wateringly high price tag. When it came to an end in 1973, its cumulative cost was estimated at $25.4 billion – and a billion dollars went a lot further in those days. You'd have to multiply that figure by at least six to get an idea of its equivalent today. Coupled with the fact that Apollo was a standalone programme that didn't really lead anywhere, this can look like an awful waste of money.

Of course, the money didn't really go to waste, or even end up entirely in the bank accounts of NASA employees. It was cycled back into the US economy in the same way that most government expenditure is. Although the Apollo spacecraft, and its enormous Saturn V launcher, were designed by NASA, their actual manufacture was contracted out to industry. Many of the giants of the US aviation industry – Boeing, North American, McDonnell Douglas and the Grumman corporation – had a share of the work.

It was easy money, too, in the form of 'cost-plus-fixed-fee' contracts. In other words, NASA agreed to pay the companies all the costs they incurred for materials and labour – whatever these ended up being – plus a flat fee on top. The latter was guaranteed profit for the company, leaving them no incentive whatsoever to minimise their expenditure. It's a great business model for the companies, and a terrible one for the US taxpayer. It's interesting to speculate how much less the Apollo programme might have cost if it had been done on firm-price contracts instead.

At the opposite extreme from Apollo was a later US mission called Lunar Prospector. Admittedly this was only a small robotic probe, which orbited the Moon rather than landing on it, but it was designed and flown on the shoestring budget of $63 million. That would still be a pretty hefty sum if it suddenly turned up in your personal bank account, but you have to remember that it was spread over hundreds of people in the ten years between 1988, when the concept was first proposed, and 1998, when it was finally launched.

Unlike any previous space mission, Lunar Prospector was essentially the brainchild of a single individual, Dr Alan Binder of the Lockheed Martin corporation. A scientist and a businessman, he brought both of these disciplines to bear on the project. On the one hand, he never lost sight of its scientific objectives – to carry out a compositional survey of the lunar surface with a particular focus on the search for water ice – while at the same time designing the spacecraft, and planning its operation, as cost-effectively as possible. As a result, Lunar Prospector was one of the slickest and most efficient space missions of the 20th century, orbiting the Moon for eighteen months and detecting around 6 billion tonnes of ice hidden in craters near the lunar poles.

In many ways, Lunar Prospector shows the way forward for private space missions far more than Apollo does. It was conceived from the start with an eye on keeping costs down, and although it had a scientific objective, that object-ive – finding frozen water – was one that will be key to any commercial exploitation of the Moon. Even the name, 'Lunar Prospector', was more suggestive of fortune-seeking private individuals than altruistic government agencies.

While he was working on the project, Binder became so disgusted by what he saw of NASA's profligate working practices that, as soon as the mission was over, he siphoned all his frustration into a huge book entitled *Lunar Prospector: Against All Odds*. Its aim, in his own words, was 'to show the American taxpayers, who foot the bill for the NASA space program, just how poorly NASA and the big aerospace com-panies manage and conduct the American space program'.

Although in some ways it can be seen as the proto-type of a private sector space mission, the fact remains that Lunar Prospector was a government-run project in all its important aspects, from funding to launch and in-flight operations. The same can be said of the first so-called 'space tourist' flights, with private citizens launched into space on what in all other respects were standard government-operated missions.

Paying Passengers

'Astronauts wanted, no experience required.' Imagine hearing that on the radio one morning, deciding to apply on the spur of the moment – and then ending up after the requisite training in an orbiting space station 360 km above the Earth. It may sound like another Arthur C. Clarke scenario, but it happened in real life to Helen Sharman.

In 1989 there were still only two countries, the United States and the Soviet Union, with spacecraft capable of launching humans into space. But over the years, scientific specialists from no fewer than eighteen other countries had flown in those spacecraft, their places paid for by their respective governments. We're talking about people from, for example, Bulgaria, Vietnam, Cuba, India and Afghanistan on Russian Soyuz flights, or Canada, Saudi Arabia, Mexico and the Netherlands on the American Shuttle.

Conspicuous by its absence from the list is the United Kingdom. The sad fact is that successive UK governments showed no interest whatsoever in joining the human space-flight club. It was left to a private consortium of British companies to set up Project Juno, with the aim of raising sufficient funds to purchase a seat on a Soyuz flight for a UK citizen. From that initiative, in 1989, came the attention-grabbing radio advert mentioned above.

Helen Sharman was just one of around 13,000 people who applied for the job, and not an obvious shoo-in either. It's embarrassing to admit in hindsight, but that was partly because of her gender. At the time, only one other woman had ever flown on a Soyuz craft. Sharman's scientific specialism didn't help a great deal either. She had a PhD in chemistry, but her job at the time – with the Mars confectionery company – was all about the flavourings in chocolate bars, which is hardly the most obvious career path to outer space. On the other hand, the media were able to refer to her as 'the girl from Mars', which was a boost for her public image if nothing else.

Despite the odds, Sharman ended up as one of four candidates selected for training in the Soviet Union, and then as primary choice for the actual mission, Soyuz TM-12. Unfortunately, as the launch date approached in May 1991, Project Juno still hadn't raised the full price of her seat, which was £7 million. However, as students of history probably know (and older readers will remember), the Soviet Union was on its last legs at that time, and needed all the international goodwill it could get. The Soviet President, Mikhail Gorbachev, agreed to waive the remaining costs, and Sharman's flight went ahead as planned.

The main purpose of Soyuz TM-12 was to ferry two new crew members to the Mir space station, a smaller predecessor to the ISS. However, Sharman herself spent just eight days on the station, before returning with the previous crew on board Soyuz TM-11. That's the same pattern we've now come to associate with 'space tourists', and it's tempting to say that Sharman was the first of these. But that's not strictly true. Although her seat was paid for with (mostly) private sector funds, it wasn't her own money. On top of that, her contract specified that she had a job to do when she was up there in space. This involved carrying out a number of experiments on behalf of her sponsors, including educational projects with British schools.

Genuine space tourism had to wait another decade. By this time, the Soviet Union was long gone, and its successor, the Russian Federation, had lost any qualms it might have had about making money capitalist-style. Eric Anderson, an American entrepreneur, formed a company called Space Adventures – effectively the first outer-space travel agency – to exploit the situation. The idea was that the company would buy seats on Soyuz flights to the ISS, which by this time had replaced Mir, on behalf of wealthy customers.

As with any start-up business, Anderson needed funds to get going. One of the people he approached in the hope of getting a loan was a financial fund manager named Dennis Tito. The latter, however, had a different idea. Rather than financing Space Adventures, he would be its first customer. He paid $20 million for his trip into space, which followed a similar pattern to Helen Sharman's. On 28 April 2001, Tito flew up on Soyuz TM-32 alongside two new ISS crew members, then spent just over a week living on the station before returning on Soyuz TM-31 with two members of the previous crew.

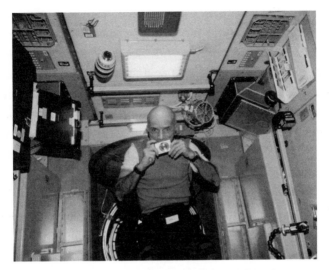

**The first self-funded space tourist,
Dennis Tito, on board the ISS in 2001.**
(NASA image)

Over the next few years, six other space tourists followed
in Tito's footsteps – one of them, software guru Charles
Simonyi, travelling not once but twice, in 2007 and 2009. To
start with, the price stayed constant at $20 million, but after
that it began to inflate at a daunting rate. Simonyi had to pay
$25 million for his first flight, and $35 million for the second. The final Space Adventures client, Guy Laliberté – the
co-founder of Cirque du Soleil – had to fork out a whopping
$40 million for his trip into space, twice the amount the first
few space tourists had paid.

After Laliberté's flight, this particular private-enterprise
route to orbit dried up (although, as we'll see in Chapter 4,

other routes will soon be opening up). The main reason for its decline was that, following the retirement of NASA's Space Shuttle in 2011, virtually every Soyuz seat was needed by ISS crew members. At the same time, that escalating price tag can't have helped. Not many people can afford $40 million, or the months of effort involved in pre-flight training, for a flight into space, no matter how keen they are. Fortunately for budding space tourists, there are several cheaper options – and we'll take a look at some of them now.

SUBORBITAL ADVENTURES

2

Is it possible to experience the thrill of space travel without having to fork out millions of dollars for a week on the ISS? The answer depends on what you mean by 'space'. From the personal experience point of view, perhaps it just means floating in a 'zero-g' environment, which is the most obvious thing we notice when we see astronauts on board the ISS. Or does it mean being high enough to see the Earth as a spherical globe against a black sky, like a planet in a sci-fi movie?

As we saw in the previous chapter, you could experience both those things by taking a short suborbital hop above the Kármán line, following in the footsteps of Alan Shepard. Two private companies, Jeff Bezos's Blue Origin and Richard Branson's Virgin Galactic, are planning to offer such trips in the near future – and at a fraction of the cost of an orbital flight. Before we look at those offerings in detail, however, there are a couple of even cheaper options to consider. If you're happy to settle for a single experience – either weightlessness or a high-altitude perspective, but not both at the same time – there are possibilities open to you right now.

Weightlessness

Everyone knows what happens when you go into space: you float, as if there's no gravity holding you down. It's a condition that goes by many names, such as 'weightlessness', 'zero-g' or – in the early days of space flight, though less so today – 'free fall'. NASA tends to use the word 'microgravity', which gives the impression that the other terms aren't strictly accurate. But they're all correct – they just emphasise different aspects of the underlying physics, as we can see if we take a closer look at how gravity works.

We generally think of gravity as the force that holds us to the surface of the Earth, but actually it's a force that exists between any two bodies having mass. The force is proportional to the two masses multiplied together, and it just happens that the Earth is the only object we ever encounter that's massive enough for us to feel its gravity.

The other factor that determines the strength of gravity is the distance to the centre of attraction – or the centre of the Earth, in our case. The force drops off with the square of this distance, so you'd expect it to be weaker at higher altitudes. At sea level, the centre of the Earth is 6,370 kilometres away, whereas all the way up at the Kármán line, a hundred kilometres higher, it's – well, it's 6,470 km away. That's hardly any difference at all, is it? The fact is that gravity is almost as strong just above the Kármán line as it is down here on the surface. That's why Alan Shepard's Mercury capsule, and the German MW 18014 test rocket, immediately started falling back to Earth after reaching apogee.

That's the objective reality, anyway, as seen by an outside observer. The subtlety lies in what happens *subjectively*, for example inside a falling Mercury capsule. It really is subtle, too, ultimately only being explained by Einstein's theory of general relativity. It turns out – and even Einstein struggled to grasp this at first – that anyone inside the capsule will be totally oblivious to the force of gravity, even though it's acting on them as strongly as ever.

This is where all those different terms come from. It's 'free fall', because the capsule is falling freely under gravity. It's 'zero-g', not because there's zero gravity, but because you're not subjectively aware of any acceleration or 'g-force'. And it's 'weightlessness', because weight is just another word for the force of gravity that we feel all the time on the surface of the Earth.

The same physics explains why astronauts are 'weightless' when they're on the ISS in Earth orbit. Despite what some journalists might tell you, this isn't because they're 'beyond the pull of Earth's gravity'. If they were, the ISS would fly off at a tangent and be lost in space forever. Instead, as we saw in the previous chapter, an orbiting object is actually in free fall – but it has just the right amount of sideways velocity that it's constantly falling *around* the curved Earth, without ever crashing back into it. That pedantic NASA term 'microgravity' doesn't mean the astronauts feel some tiny residual effect of Earth's gravity either, because they don't. What they do feel are minuscule g-forces from other sources, including the small but non-zero gravitational pull of the space station itself.

For a body to be in 'free fall', it doesn't actually have to be falling downwards. Strange as it may sound, it can be 'falling' upwards too. To see why, let's think back to Alan Shepard's brief jaunt into space in 1961. At launch, he was pushed upwards by the Redstone rocket engine, and he certainly wasn't in free fall then. In fact, the upward acceleration of the rocket added to the g-forces, with Shepard subjected to more than 6g – that is, six times the force of gravity at the Earth's surface – by the time the Redstone's engine had built up to maximum thrust.

Then, 142 seconds after launch, the engine shut down and the Mercury capsule, with Shepard inside, separated from the booster rocket. At this point it was just 60 km above the Earth, still well inside the atmosphere, and entirely at the mercy of gravity. But its accumulated momentum continued carrying it upwards, on a parabolic trajectory, for a further two and a half minutes, until it reached apogee at around 188 km. It was during this time that Shepard was effectively 'falling upwards', since the capsule's motion was determined entirely by gravity. This continued to be the case after reaching the top of the parabola and starting the more obvious fall back down to Earth.

Shepard was weightless throughout the parabolic part of the flight, when the capsule's motion was dictated solely by gravity and its own momentum. This was the case from the moment the engine cut out, until atmospheric drag on the descending capsule became strong enough to have an appreciable braking effect. The take-home message here is that his experience of weightlessness had nothing to do with the altitude he was flying at, or whether he was technically 'in space' or not. It was all down to the trajectory of the capsule he was travelling in.

There are various ways to achieve a weightless state without travelling into space. Einstein's preferred method was to plunge down a lift shaft after the elevator cable snaps. But that's always going to end badly, and fortunately he only ever indulged in it as a purely cerebral 'thought experiment'. A much safer approach is to simulate a Shepard-style parabolic trajectory inside a specially designed aircraft. Originally developed in the late 1950s as a way to train astronauts, these aircraft are commonly known by the nickname 'vomit comets', due to the physiological effect that weightlessness has on some participants.

As with so many of the topics in this book, it was several decades before it dawned on anyone that there might be a private-sector market for vomit comets. In the end, it was one of the great pioneers of space tourism, the entrepreneur Peter Diamandis, who seized the opportunity. He set up the Zero-G corporation, which operates a converted airliner kitted out like one of NASA's astronaut training planes. A typical flight includes up to eighteen parabolic segments, each providing 20–40 seconds of weightlessness. They're available through the usual outer space travel agency, Eric Anderson's Space Adventures, for around $5,000 per person

– and in practice, flights are often block-booked by a single company as a reward for a group of employees – so it really is the easiest way for the average person to get a pseudo-space experience.

Zero-G started operating in 2004, half a century after the death of Einstein, the person who started us thinking about weightlessness with his musings about falling elevators. Einstein's best-known successor in the field of general relativity, the late Stephen Hawking, became a Zero-G customer himself when he took a flight in April 2007. Given Hawking's extreme physical disability, which confined him to a wheelchair when he was earthbound, the experience of weightlessness was particularly liberating. 'People who know me well say that my smile was the biggest they'd ever seen,' he remarked afterwards.

Stephen Hawking experiencing a Zero-G flight in 2007.

(NASA image)

So it's possible to experience the *feel* of outer space for a reasonably affordable price. But what about the *view* from outer space?

The Edge of Space

Many people will remember the dramatic day in October 2012 when the daredevil adventurer Felix Baumgartner ascended by helium balloon to a record-breaking altitude of 39 kilometres over Roswell in New Mexico – and then jumped all the way back to Earth, at one point exceeding the speed of sound before his parachute finally opened. Close to 10 million viewers around the world watched the 'space jump' live on YouTube – a record audience that still stood at the end of the decade.

Of course, 39 km isn't really space. It's less than two-fifths of the way to the Kármán line, and by definition a balloon can't reach outer space, because it needs a certain amount of atmosphere to support it. But the fact remains that it *looked* like space. The ascent had the feel of a space mission, and Baumgartner was kitted out in a spacesuit that any astronaut would be proud of. The view of Earth, just before he jumped, hardly looked any different to the view from the ISS.

Baumgartner's altitude of 39 km falls in a kind of no man's land, well below the Kármán line but well above the altitude at which most commercial and military aircraft fly, which rarely get higher than 20 km. It's a region that's sometimes referred to as 'proto-space', and it undoubtedly has tourist potential. Not many people would want to emulate Baumgartner by jumping from that height, but they might happily travel up there in a balloon, as long as they knew it was going to come safely back down to Earth again. You can even imagine huge floating hotels, in which people could spend days or weeks cruising in 'almost space'.

It's possible that a first step in this direction may happen before too long. In June 2020, a Florida-based company called Space Perspective announced plans to offer balloon trips up to 30 km altitude. Their proposed balloon, called Spaceship Neptune, would house passengers inside a pressurised capsule, avoiding the need for Baumgartner-style spacesuits. Another novelty, compared to most other high-altitude balloons, is that Neptune would use hydrogen gas instead of helium. From a pure physics point of view this makes a lot of sense, because hydrogen has a lower density than helium. As we saw in the previous chapter, balloons rise through the atmosphere because they weigh less than the air they displace, so a hydrogen balloon will always rise further than a helium balloon of the same size.

Hydrogen has other benefits over helium, not least that it is much cheaper and easier to produce. Its most obvious disadvantage, as most people will be aware, is that it's extremely flammable. Hydrogen plus oxygen plus any kind of ignition source equals a ferocious fire – such as the one that destroyed the hydrogen-filled Hindenburg airship in 1937. These days, however, well-designed hydrogen balloons are essentially as safe as the helium variety, and they're steadily making a comeback. The prospective operators of Spaceship Neptune clearly believe in their effectiveness, anyway. Here's what their initial press release had to say:

> Neptune takes up to eight passengers, called 'explorers', on a six-hour journey to the edge of space and safely back. It will carry people on a two-hour gentle ascent above 99 per cent of the Earth's atmosphere to 100,000 feet, where it cruises above the Earth for up to two hours allowing passengers to share their experience via social media and with their fellow explorers. Neptune then makes a two-hour descent under the balloon and splashes down, where a ship retrieves the passengers, the capsule and the balloon.

That's all still to come, but if you want a rather more adrenaline-pumping way to reach 'the edge of space', there's one that's been available to those who could afford it for some time. One of several offerings from the Moscow-based company RusAdventures involves high-altitude flights in a MiG-25 'Foxbat' jet fighter. Designed during the Cold War with the specific aim of intercepting high-flying spyplanes, this still holds the altitude record for a conventional jet aircraft of 37 km. RusAdventures offers customers a more modest 25 km, but that's still high enough to see the curvature of the Earth against the blackness of space. And you can be pretty confident that the only people higher than you are on board the ISS.

Hyperbolically referred to as 'Flight to the Stars', the Foxbat package doesn't come cheap, at $20,800 per person for a four- or five-hour flight. But if you want to see the Earth from (almost) space, and you want to do it now rather than waiting for some vague time in the future, then it's your only option.

Spaceplanes

As we saw earlier on, the only way to get into *real* space, above the Kármán line, is with a rocket. Jet aircraft and balloons simply won't work at that altitude. But a rocket is a type of engine, not a type of vehicle – and it doesn't have to drive the kind of vehicle that the word 'rocket' might immediately call to mind. That's basically a long cylinder launched vertically from ground level – anything from the Redstone that pushed Alan Shepard over the Kármán line to the Saturn V that took Apollo 11 to the Moon.

In the same decade in which those two flights took place, the 1960s, NASA also experimented with a completely different way to get into space. This took the form of the X-15 – a winged aircraft that flew horizontally like a jet, but was powered by a rocket engine that allowed it to reach much higher speeds and altitudes. It was launched from beneath the wing of a converted B-52 bomber at an altitude of around 13 km, not much higher than a commercial airliner. Its rocket engine then ignited for about a minute and a half, pushing the X-15's speed up to 6,800 kph.

If the pilot chose to point the aircraft up at a steep angle (which wasn't done on every test flight), then the plane could reach extremely high altitudes. As with Alan Shepard's flight, this essentially involved an unpowered parabolic trajectory, initially being carried upwards by sheer momentum, then dipping down and heading back to Earth. Although the X-15 had wings, these didn't provide any additional lift at high altitude because the air was too thin. As a consequence, the pilot was in free fall all this time, just as Shepard had been, and experienced weightlessness in the same way. It was only when the plane dropped down into thicker air that the wings started to provide lift and the pilot's sensation of weight returned. The plane then glided back to a landing on the ground.

The primary aim of the X-15 project was to test the concept of a super-fast rocket plane, not to reach space or to break altitude records. So, in the event, only two X-15 flights actually breached the Kármán line, both flown by test pilot Joe Walker. On 19 July 1963 he reached an altitude of 106 km, followed by 108 km on his next flight on 22 August of the same year. Significantly, both flights involved the same vehicle, which highlights one of the key features of the X-15: it was reusable. At the time, the technology required to make a vertically launched rocket reusable still lay decades in the future.

From NASA's point of view, its reusability was the only real advantage the X-15 had over a conventional space launcher. The flight was too brief, and the payload capacity too small, to make it a practical solution to any problem they were interested in. On the other hand, from a space tourism perspective – which needless to say never crossed anyone's mind at the time – it's a different matter altogether. The fact that it's reusable keeps costs down, making it more likely that a profitable business could be built around it. The horizontal launch makes it a more comfortable experience for those who haven't had astronaut training. Yet the flight still ticks all the space tourism boxes – it gives passengers a spectacular view of the Earth, it passes the symbolically important Kármán line, and it provides an extended period of weightlessness – at least ten minutes in the case of Joe Walker's flights.

So why didn't private companies queue up to follow in the X-15's footsteps and build their own spaceplanes? One obstacle seems to have been the ingrained belief, mentioned in the first chapter, that space was the exclusive province of governments. Another was the view that any use of space, other than for scientific or military purposes, had no place outside daydreams and science fiction. In the early 1990s, Peter Diamandis tried to persuade the International Space University in Strasbourg to put spaceplanes on the curriculum, but the senior faculty told him that such things were pie in the sky and had no place in serious academia.

Undeterred, Diamandis decided to set the world a challenge. The first privately built and operated spaceplane to carry a human above the Kármán line – not once, but twice in the space of two weeks – would win a prize of $10 million. Called the X Prize in honour of the X-15, it was first announced in 1998 at the Smithsonian Air and Space Museum in Washington DC. The only minor problem was that Diamandis didn't have the prize money yet.

He went looking for sponsors, and found the Ansari family from Texas, who shared his enthusiasm for the privatisation of space. In fact, Anousheh Ansari later became the first female Space Adventures customer, spending eight days on the ISS in September 2006. The Ansaris were prepared to put up the full amount of the prize money, but in the event they thought up a better scheme.

They realised that almost everyone in the world, except Peter Diamandis and themselves, was convinced the prize money would never be claimed. So they took out an insurance policy *against* the eventuality of having to pay the $10 million. The insurance company charged them a premium estimated at $1 million, in the smug belief that the Ansaris were suckers. But they weren't, because that million was all they ever had to pay. When the prize was won, in October 2004, it was the insurance company that had to fork out $10 million for the winner.

That winner, one of over a dozen serious contenders for the prize, was a small company with the peculiar name of Scaled Composites. If that sounds a little uninspiring, they made up for it with their spaceplane, which was grandiosely dubbed SpaceShipOne. In fact, it was a much more modest affair than the X-15 – smaller, less powerful and with a lower top speed. But in other ways it was a more sophisticated design, using a hybrid solid/liquid-fuelled rocket engine and a unique 'feathering' system which flips the wing forward for additional braking during descent. Most crucially of all, SpaceShipOne had a 'flight ceiling' – maximum altitude, in other words – of over a hundred kilometres, just like the X-15.

SpaceShipOne was air-launched in the same way as the X-15, but rather than using an existing aircraft such as the B-52, Scaled Composites designed a purpose-built carrier aircraft called White Knight. This carried the spaceplane to an altitude of 14 km, where it was released and its rocket engine fired up. On the first of its two prize-winning flights on 29 September 2004, the rocket burned for 80 seconds to put SpaceShipOne on course for an apogee at 103 km, just over the Kármán line. The feathering system then kicked in to bring it down in a graceful glide back to the ground. The total flight, from air release to landing, was around 24 minutes.

That first flight was piloted by Mike Melvill. Five days later, the same aircraft was taken up again, this time to 112 km, by a second pilot, Brian Binnie. It was the combination of these two flights, in rapid succession, that won the X Prize for Scaled Composites. What so many people, for so long, had maintained would never happen – a private flight into space – had finally been achieved.

One person who was impressed by the feat was Richard Branson, the billionaire owner of, among other things, the Virgin Atlantic airline company. Almost immediately, he commissioned Scaled Composites to build a larger version of SpaceShipOne, capable of carrying six passengers and two crew members on the same kind of suborbital flight above the Kármán line. The flights would be offered through a company that Branson called – wittily, if inaccurately from an astronomical point of view – Virgin Galactic. Before the end of 2004, the year that Scaled Composites won the X Prize,

Virgin Galactic was already a registered business, with Branson confidently promising passenger-carrying flights within three years.

Unfortunately, this estimate proved over-optimistic – by a long way. Part of the problem was that the new space-plane, inevitably named SpaceShipTwo, had to be a lot more complex than the initial test vehicle. It was bigger, for one thing – the size of a small business jet – and all the novel technology of the prototype, including the hybrid engine and wing-feathering arrangement, had to be scaled up accordingly. Similarly, the carrier vehicle, WhiteKnightTwo, also had to be a much larger aircraft. It ended up as a twin-fuselage

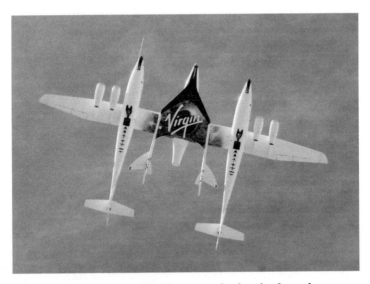

Virgin's SpaceShipTwo, attached to its launch craft WhiteKnightTwo, as seen from below.

design with a total wingspan of over 40 metres. In the end, SpaceShipTwo's first powered flight had to wait until April 2013 – six years after Branson's first customers had been expecting to fly into space on it.

The first SpaceShipTwo to come off the production line was called *Enterprise*, after Captain Kirk's starship in *Star Trek*.* After its first successful flight, Branson – still as optimistic as ever – revised his estimate of the start of commercial operations to 2014. Sadly, even that wasn't going to happen. The test flights didn't go as well as hoped, with *Enterprise* failing to reach its design spec both in terms of speed and altitude. Then on the fourth test, on 31 October 2014, the spaceplane disintegrated soon after igniting its rocket engine. One of the two pilots parachuted to safety, but the other was tragically killed. The problem was that the feathering system, effectively a giant air brake, had been deployed too early in the flight.

As well as being disastrous in human terms, the crash was a huge setback for Virgin Galactic. The aforementioned 'production line' wasn't a fast-moving one, and the second model, *Unity*, didn't roll out until February 2016. Like its predecessor, *Unity* has struggled to meet its design spec in test flights, such as the much-publicised one in July 2021 which took Branson himself to an altitude of just 86 km,

---

* William Shatner, the actor who played Kirk, was one of many celebrities approached by Branson with a view to buying Virgin Galactic tickets. Shatner's response was: 'Hey, you pay me and I'll go up. I'll risk my life for a large sum of money.' Branson didn't get back to him.

well below the Kármán line. In any case, the situation has changed dramatically since Virgin Galactic was founded in 2004. Then, Branson's suborbital offering was the only one in town. Now he has a competitor, and one that has already caught up to him, in the form of fellow billionaire Jeff Bezos.

Reusable Rockets

Bezos is best known as the founder of the e-commerce giant Amazon – and as one of the richest people in history, with a net worth in the region of $200 billion. This puts him in an unrivalled position to invest in risky but potentially game-changing sidelines such as his Blue Origin space launch company. In essence, this is a direct competitor to Virgin Galactic – offering space tourists brief flights above the Kármán line, complete with several minutes of weightlessness and spectacular high-altitude views of Earth – but it goes about it in a different, and actually much simpler, way.

If Richard Branson's SpaceShipTwo can be thought of as a private-sector reboot of NASA's X-15 rocket plane, then Bezos' approach is a homage to Alan Shepard's flight atop the vertically launched Redstone. To make the connection completely explicit, Blue Origin even named their rocket 'New Shepard'.

There really isn't a simpler way to get into space than launching vertically upwards from the ground, and then coming back down in a small capsule that detaches from the launcher and descends by parachute. There's no need for any of the things that make SpaceShipTwo so complicated. You don't need a carrier aircraft or a hybrid engine or a variable-geometry wing – or even a pilot, since the whole flight can be controlled by a computer. The downside is that, while spaceplanes are reusable, a traditional rocket is single-use. This means that a regular launch schedule becomes very expensive very quickly.

At any rate, that was the situation when SpaceShipTwo's predecessor won the X Prize back in 2004, and it's why Virgin Galactic's approach looked like the only commercially viable one. But is that really true? Is there any fundamental reason why a rocket that takes off vertically can't come back and land the same way? It turns out the answer is no. It's a big technical challenge, but not an impossible one.

Rocket thrust can be used to slow down a descending rocket just as well as it can accelerate an ascending one. The difficulty isn't with the rocket engine itself, so much as the electronic systems that keep it stable and allow it to navigate to a precise touchdown at the desired spot. That's something that, for many decades, only ever happened in sci-fi movies – but now it's become quite a commonplace sight, particularly with the Falcon 9 boosters that SpaceX uses to launch satellites and crew-carrying Dragon capsules into orbit.

The first occasion that a Falcon 9 made a successful landing after a trip into space was on 21 December 2015 – but it wasn't a world first. The previous month, on 23 November, New Shepard had beaten it into the record books. That particular flight was also the first time a New Shepard had flown past the Kármán line, with an apogee just 535 metres above the magic 100 kilometres. The same rocket flew again four more times before it was retired.

Of course, a SpaceX-style orbital flight is a much greater technical challenge than a suborbital one – and a more immediately profitable one, with customers like NASA waiting in the wings. So New Shepard's achievement was eclipsed rather quickly. But from Blue Origin's point of view, it was a huge step forward, because it meant they could compete with Virgin Galactic using a system that was both simpler and cheaper to operate than SpaceShipTwo.

There are two parts to New Shepard: the booster rocket, which Blue Origin calls a 'propulsion module', and the pressurised crew capsule. The latter is spacious enough to carry six passengers in comfort, and let them float around during

The New Shepard launcher and capsule land separately – the first under its own power, the latter by parachute.

(NASA images)

the period of weightlessness. The capsule also has what the company proudly describes as 'the largest windows in space-flight history', making up a third of the capsule's exterior, and allowing passengers an unprecedented view of the Earth from space. As with Alan Shepard's flight, the capsule separates from the propulsion module after the boost phase and returns to Earth by parachute, while the booster lands under its own power.

New Shepard first flew in its operational configuration in December 2017, after which the same hardware was reused on half a dozen further flights, albeit uncrewed ones. In contrast to Virgin Galactic's struggles with SpaceShipTwo, all these flights easily came up to New Shepard's intended design spec, with the first crewed test flight, in July 2021, taking Bezos and three other passengers to 105 km altitude, comfortably above the Kármán line.

If you've ever seen a video of a New Shepard flight, you'll have been struck by how gentle and decorous the whole thing is compared to an orbital spaceflight. For one thing, it's a surprisingly small vehicle, just 18 metres high compared to Falcon 9's 70 metres. And while the latter can reach an orbital speed of 28,000 kph, New Shepard's maximum is a much more leisurely 3,500 kph. That's only a little over one and a half times the cruising speed of the Concorde supersonic airliner, which was 2,200 kph. The upshot for prospective passengers is that they'll be spared the sustained high-g accelerations that professional astronauts have to contend with.

That's true of Virgin Galactic's paying passengers too, who will also have a pretty comfortable ride into space. But just like Blue Origin, at the time of writing (2021) we still have to insert that word 'will' – future tense. Which of the two rivals will be first to start regular commercial operations isn't an easy call to make. Richard Branson's company had a massive head start, but Bezos's is making faster progress, with fewer setbacks. To confuse matters further, Branson has a history of announcing over-optimistic timescales, while Bezos has always been tight-lipped about them.

There's another difference between the two companies. While Blue Origin hasn't decided on a ticket price yet, Virgin Galactic has been taking deposits from prospective customers for years. The original ticket price was $200,000, which rose to $250,000 in 2013. By the following year, some 700 people had already booked flights – and only 20 of them requested refunds after the *Enterprise* accident. Nevertheless, it's hard to believe that the remainder aren't getting pretty frustrated and impatient by now.

The plain fact is that, as a business, suborbital space tourism hasn't got off the ground yet. No one can be sure if it will even turn out to be a viable way of making money. But Virgin and Blue Origin aren't putting all their eggs in one basket. Both companies are planning to supplement their space activities by breaking into a better established, and demonstrably profitable, market: orbital launch services. Richard Branson has set up yet another company, Virgin Orbit, for this purpose. The aim is to put small satellites into orbit using a rocket fired, missile-style, from under the wing of an aircraft. And Blue Origin is working on a successor to New Shepard called New Glenn – after John Glenn, the first

INTO ORBIT   3

If you were around in 1967, and happened to switch on the television on 25 June of that year, you might have caught a live broadcast called *Our World*. That's true whether you live in the United Kingdom or the United States, or in Japan, Australia, Mexico, Tunisia or any of two dozen countries around the world. Fourteen of those countries contributed live material to the broadcast, which famously culminated in the world premiere of 'All You Need Is Love' by the Beatles.

At this point you may be thinking 'So what?' After all, live TV linkups happen all the time, and while *Our World* may have been one of the cultural high points of the 1960s, what's it doing in a book about the commercialisation of outer space? The answer is that, for its time, it was an astounding technical achievement which would have been unthinkable before the advent of commercial communication satellites. The first of these, Early Bird, was launched into a geosynchronous orbit over the Atlantic in 1965. Together with three subsequent satellites, at other locations around the globe, it was what made the *Our World* broadcast possible.

One of the ironies of the space age is that, as soon as a technology becomes really useful, people forget that it has anything to do with space travel. They'll tell you that space is a wasteful, elitist activity that has no relevance to ordinary people, while using the word 'satellite' in everyday language without even thinking what it means. Yet many of us would be lost, literally, if we didn't have satellite navigation in our cars, while others would be lost in a more figurative sense without satellite TV to watch in the evenings. Then there's the satellite imagery on Google Earth and Apple Maps, and in TV weather bulletins. The fact is, modern life as we know it simply wouldn't be possible without all that hardware we've put in orbit.

In 2017, before SpaceX began to deploy its network of Starlink broadband satellites, there were around 1,200 operational satellites – a number that will be surpassed enormously by the time the Starlink constellation is finished. Even on the 2017 statistics, commercial satellites outnumbered government-owned ones. Only about a fifth of the total were devoted to purely scientific or other research activities, while another fifth were for other government activities, including military surveillance. The next fifth – a mixture of government and privately owned satellites – were for Earth observation and GPS-style navigation. That left a whole two-fifths, and growing, for commercial communications.

When the 40-kilogram Early Bird was launched in 1965, it had a capacity of 240 audio channels or a single, low-resolution black-and-white video channel. That exceeded anything that had been possible over intercontinental distances with a wired network, but it was still only the beginning. The most powerful of its modern-day successors weigh six tonnes or more, and are capable of handling 25,000 TV channels or a whopping 25 million audio channels. Nevertheless, such satellites share one key feature with Early Bird: they occupy the same kind of 'geosynchronous' orbit. It's worth taking a closer look at exactly what that means.

GEO and LEO

If you've ever watched a satellite moving across the night sky, you may have been struck by how quickly it's travelling. They're usually only visible for a minute or so before they vanish over the horizon. The same is true of the ISS – overhead passes of which can be quite spectacular, owing to its size and brightness – and of SpaceX's ever-growing constellation of Starlink satellites. On the other hand, if you own a satellite TV dish you'll know that it isn't supposed to move. If you've ever had to go up and adjust it after a storm, you'll be aware just how precisely it has to stare at a specific spot in the sky. Like an old-fashioned terrestrial TV mast, the satellite remains in a fixed spot relative to your house.

So what's going on? Why do some satellites move across the sky so quickly, while others don't move at all? The answer all comes down to altitude. The higher a satellite is, the slower it travels in its orbit around the Earth. The ones that are close enough for us to see them in the night sky are in what's called 'low Earth orbit', or LEO, just a few hundred kilometres above the Earth's surface. That's true of the great majority of satellites, and of all crewed missions in Earth orbit – including the ISS, which orbits at around 400 km altitude. The main exceptions to this rule are satellites which, for one reason or another, need to remain fixed over the same spot on the Earth's surface. This includes, for example, broadcasting satellites and weather satellites.

A satellite in a circular orbit at a particular altitude has absolutely no freedom to choose the speed it travels at. It's determined by Newton's law of gravity – or, if you want to be really pedantic, by Einstein's theory of general relativity. As daunting as that sounds, it just means that if you plug the desired speed into a standard formula, out will come the necessary altitude. But the same isn't true for the speed the planet itself rotates at. There isn't any fundamental physics governing this, and to all intents and purposes it's a random number. In the case of the Earth, the rotation speed just happens to be one revolution in 24 hours.

So that's all we have to know in order to set up a geosynchronous Earth orbit (GEO). We take the standard formula and insert a speed of one orbit per 24 hours, and it tells us the altitude we need to put our satellite at. The answer turns out to be a very big number in comparison to LEO – around 36,000 km. It's not surprising that people only put satellites in GEO when they really have to.

The first chapter opened with some of Arthur C. Clarke's far-sighted speculations on the commercial uses of space. The

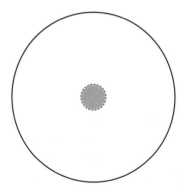

A scale diagram showing Earth (the shaded circle at the centre), the 'low Earth orbit' of the ISS (dotted circle) and a geosynchronous orbit (solid outer circle).

geosynchronous orbit is probably the most famous of these – in fact it's sometimes referred to as a 'Clarke orbit' in his honour. He wasn't the first person to realise that a satellite orbiting 36,000 km above the equator would always remain over the same spot on Earth, but he was the first to push the idea of a network of communication relays at that altitude.

That was in a non-fiction article he sold to *Wireless World* magazine in 1945. Twenty years later, when geosynchronous satellites had become a reality, he came to regret not having tried to patent the idea instead. As he wrote in his 1966 book *Voices from the Sky*, 'It is with somewhat mixed feelings that I can claim to have originated one of the most commercially valuable ideas of the 20th century, and to have sold it for just 40 dollars.'

Another point Clarke makes in the same book is how rapidly technology advanced between the end of the Second

World War and the 1960s. When he wrote that original *Wireless World* article, electronic hardware used large, fragile vacuum tubes* to control the flow of electrons – and these had a habit of burning out with depressing regularity. The idea of a small, self-contained satellite like Early Bird would have been unthinkable. As Clarke says:

> I envisaged my 'extraterrestrial relays' as large structures with their own maintenance and operating crews, but miniaturisation, and above all the invention of the transistor, made it possible for tiny robots to do the work of inhabited space stations.

The advantage of a geosynchronous satellite is that, from the point of view of a homeowner, it's functionally identical to a traditional TV mast. You can have a small, highly directional receiving antenna that, once it's been properly installed, doesn't need any smart tracking or processing software. It picks up the signal from the satellite – which, because it's a relatively narrow beam, doesn't have to be a particularly strong signal – and then amplifies it before sending it to your TV.

This is all possible because the satellite is in that special orbit, perpetually hovering over the same spot on the rotating Earth. But recalling just how far away the satellite is

---

* If you're too young to remember these, picture an old-style incandescent light bulb with various metal plates and electrodes inside it. If you're too young to remember incandescent light bulbs, never mind.

– 36,000 km – we can see a couple of disadvantages, too. The first is getting it up there. As mentioned earlier, large communication satellites can weigh several tonnes, and getting them to GEO – as opposed to LEO, just a few hundred kilometres up – uses a lot of rocket fuel. SpaceX's Falcon 9 can carry a 22-tonne payload to LEO, but only eight tonnes to GEO.*

The other problem is that, if we want a given signal strength at the receiver on Earth, the actual power transmitted by the satellite has to be bigger if it's further away. That's not because the signal gets attenuated as it travels through space (which it doesn't), but because it spreads out like the beam of a flashlight. Even if the beam were very narrow to start with, it will be diffused out over a much wider area by the time it reaches Earth. If we want two-way communication, the same applies to the transmitter on the ground – it has to be far more powerful in order to reach a geostationary satellite.

Weighing up the pros and cons, it turns out that GEO is the best option for things like direct-to-the-home TV broadcasting, where all that's needed on the ground is a unidirectional receiving antenna pointing straight at the satellite. GEO also works well for high-capacity two-way communications involving permanent ground stations with large, fixed-dish antennas. On the other hand, it's a non-starter when it comes to two-way communications between private individuals. They're just not going to have the time or resources to set up a powerful transmitting dish,

---

* Geostationary satellites are generally launched into low preliminary orbits, before being boosted to higher altitude.

and line it up in exactly the right direction, every time they want to talk to someone.

In spite of this, satellite phones do exist. They get round the signal strength and aiming problems by using satellites that are much closer to Earth, in LEO rather than GEO. This approach was pioneered by the Iridium corporation in the late 1990s. It won't work with just a single LEO satellite, which would be moving so quickly it might have disappeared over the horizon before you finished your call. Instead, Iridium uses an ensemble of 66 satellites working in concert with each other.

The jargon term for a satellite ensemble of this kind is 'constellation' – which is a little misleading, because the literal meaning of constellation is a fixed pattern of stars in the sky. But the satellites in a constellation aren't in fixed positions relative to each other. They're all moving around the Earth in different orbits, so as to give a roughly uniform coverage over the globe.

Because Iridium satellites are so close, less than 800 km above the Earth, users on the ground don't need to fuss with directional antennas. The satellites transmit sufficient power that an omnidirectional Iridium handset can easily pick it up, and then produce a strong enough signal of its own to talk back to the satellite. The clever part, which is all in the software, is making sure the phone uses whatever satellite is closest at a particular moment, and then – because it's travelling so fast it will only be above the horizon for about seven minutes – seamlessly switches to the next satellite at the appropriate moment.

As ingenious as Iridium phones are, there isn't a huge market for them. The fact is, they don't offer much more than ordinary mobile phones, at much higher cost. They only come into their own when you're outside the range of the mobile network – and for most people these days, that's almost never. On the other hand, the same basic idea of a 'satellite constellation' has found a much more profitable niche in recent years, and that's in the context of broadband internet services.

It's possible to use GEO satellites as two-way internet relays – the American Viasat corporation, for example, does just that – but they're not an ideal solution. Customers on the ground need large antennas, significantly bigger than a satellite TV dish, in order to transmit a strong enough signal back to the satellite. This makes it an expensive business when compared to other broadband options. On the other hand, a LEO constellation offers a much more viable way to put the internet into orbit.

Several such constellations have been proposed, or even started, by companies such as Amazon and OneWeb – the latter now partially owned by the UK government. As with so many things in the space business, however, it's Elon Musk's SpaceX that is taking the lead in this particular market, with its Starlink mega-constellation. Using mass-produced user terminals that cost just $200 or so, this aims to be competitive in comparison to existing broadband services, both in densely populated urban areas and in more out-of-the-way places that may currently have no broadband option at all.

The Starlink satellites themselves are much smaller than their geostationary counterparts – just 227 kg each – and they orbit at altitudes of around 550 km. That's nothing very special so far. If you want an eye-popping statistic, you have to look at the total number of satellites in the constellation. When it's finished, around the year 2027, that will be – wait for it – near enough to 12,000 satellites. If you remember that, at the end of 2017, there were only around 1,200 operational satellites in total, that's a breathtaking figure. The total number of Starlink satellites will vastly exceed the number of non-Starlink satellites.

From the user's point of view, a Starlink terminal is hardly any more complicated than the broadband router they're currently using. The obvious visible difference is that it has an antenna, which Musk says 'looks like a thin, flat, round UFO on a stick'. This may sound rather like a conventional satellite dish (since a UFO, as everyone knows, looks like a saucer, and a saucer looks much like a dish), but actually it's a lot more sophisticated internally.

The Starlink terminal employs a 'phased array' antenna – something that has been the norm in military applications since the 1970s, but until recently was too technically complex, and too expensive, to produce for the civilian market. Essentially it uses a flat array of sensor elements which can be steered electronically, in the blink of an eye, rather than having to rotate physically. Unlike a conventional antenna, it doesn't have to be precisely aligned in the correct direction.

This makes the setting-up process trivially simple – a selling point that Musk highlighted in a tweet in January 2020:

Instructions are simply:

– Plug in socket

– Point at sky

These instructions work in either order. No training required.

Musk's biggest problem at the moment is getting all those thousands of satellites up into orbit. SpaceX's Falcon 9 rocket is capable of launching 60 Starlink satellites at a time, and it can do that at an impressively rapid tempo. In the course of 2020, for example, there were no fewer than sixteen Starlink launches, totalling 960 satellites in all.

Unlike most uncrewed space missions these days, the Starlink launches often get a prominent mention in the mainstream media. This is great news for SpaceX's publicity team, but that's not the reason for drawing the public's attention to them. It's because the newly launched satellites can make a spectacular sight when you see them trailing across the night sky. Immediately after deployment, they're still relatively close together, and it's possible to watch one after the other appearing and disappearing. Even after they've settled into their final orbits, there are going to be an awful lot of them. By the time all 12,000 have been launched, they'll easily outnumber all the visible stars in the sky.

It will undoubtedly change our view of the night sky – making it more dynamic and interesting for most of us, but potentially much more frustrating for astronomers. Since the latter like to take long-exposure photographs of very dim, distant objects, there's the risk these will be 'photobombed' by streaks of light caused by Starlink satellites whizzing past.

But the problem isn't as bad as it's sometimes made out to be. Satellites don't produce any light of their own, and all we actually see is sunlight reflected off them. So they're invisible in the middle of the night, when the Sun is right over on the opposite side of the planet. If we see a satellite against a dark sky, it's because the Sun is below our horizon but not the satellite's (the horizon is further away at higher altitude). This situation, however, can only occur in the hour or so just after sunset, or just before sunrise.

There's another problem caused by mega-constellations such as Starlink, which is potentially much more serious. This is the risk of cluttering up LEO space to the point where satellites start to collide with each other. Collisions aren't going to happen all the time, because satellite operators know where all their functioning satellites are, and can move them to different orbits if it looks like a crash is imminent. Similarly, if a satellite is still under their control when it comes to the end of its working life, they can deliberately 'deorbit' it so that it burns up harmlessly in the Earth's atmosphere.

The real problem is with 'dead' satellites, which are no longer responding to commands from ground control. The number of these is increasing all the time, along with other items of space junk such as payload shrouds and rocket boosters, which were used during satellite deployment but are now derelict. The space community is going to have to find some way to reduce this clutter if Starlink, and the competing constellations that are bound to follow, is to operate safely.

Sweeping up Space Clutter

The risk of satellites colliding with each other is an increasingly worrying problem for the space sector. If a collision does occur, the numerous fragments resulting from it will mean that much more space junk moving on uncontrolled orbits. In turn, this makes subsequent collisions even more likely – a snowballing process known as the Kessler syndrome.

Dealing with unwanted pieces of space hardware isn't easy. They can't simply be 'destroyed' in situ, by blowing them up for example, because the resulting debris would make the problem worse rather than better. Instead, the best solution is to make a defunct satellite fall out of orbit and burn up in the atmosphere. Called Active Debris Removal (ADR), this involves using another spacecraft to rendezvous with the offending object, capture it by some means, and then cause it to deorbit in the desired way.

An operational ADR system is still years in the future, but a small-scale prototype, the University of Surrey's RemoveDEBRIS, was launched in 2018. This demonstrated a number of ADR technologies, including the capture of small test objects using a net and a harpoon.

As with the 'UFO on a stick' user terminals, the Starlink satellites are mass-produced, at a rate of six per day. This means that SpaceX can achieve genuine economies of scale – another first for them, which the space sector has never come close to in the past. The hope is that, when the system is finally up and running, it will pay for itself very quickly. Elon Musk has predicted annual revenues of $30 billion per year by 2025.

As well as pleasing shareholders, some of this money will be ploughed back into other parts of SpaceX's business, such as Falcon 9 launches for third-party customers. Anything it can do to make these more affordable will be welcome news, because, as we're about to see, the space launch business is a highly competitive one.

Commercial Launch Services

For a long time, governments had a complete monopoly on space launch services, and they didn't come cheap. Monopolies rarely do – not necessarily because the people offering them are greedy, but because there's no incentive to find ways of cutting costs. That's true both of the government agencies, such as NASA, that operated the services, and the giant aerospace companies that made the launch vehicles. With zero-risk 'cost-plus-fee' contracts, the latter had no reason to make cheaper rockets.

That's all changed now, and private companies are vying with each other to provide competitively priced launch services, not just for commercial customers but for government ones too. The big breakthrough came with the launch of SpaceX's two-stage Falcon 1 rocket on 28 September 2008. It was the first private-sector space launch in history, with the Falcon's second stage entering orbit together with a small dummy payload. It was an impressive demonstration to show potential customers what was possible – but it was just the start.

Falcon 1 was really only a test vehicle, paving the way for SpaceX's workhorse, Falcon 9. The latter is bigger in every respect. At 70 metres in length, it's three and a half times the size of Falcon 1, with nine first-stage engines in place of the single one on Falcon 1 (that's where the '1' and '9' in their names come from). Together with improvements to the engines, this means Falcon 9 is capable of producing sixteen times as much thrust as its predecessor. That's enough to put a payload of 22 tonnes into low Earth orbit (as already mentioned, it can launch 60 Starlink satellites in one go). The first flight of a Falcon 9 took place on 4 June 2010, and by November 2020 it had completed its 100th flight.

But that doesn't mean that 100 Falcon 9 rockets have gone into space. Not 100 first-stage boosters, anyway, because these were designed from the start to be reusable. As described in the previous chapter, bringing a rocket stage safely back down to Earth, and landing it vertically, isn't a simple task. It took SpaceX a long time to perfect the system, and as we've seen, Blue Origin's New Shepard just beat them into the record books. It achieved its first propulsive landing in November 2015, with SpaceX duplicating the feat a month later when a Falcon 9 booster came back and landed a short distance from the launchpad at Cape Canaveral.

When it came to actually reusing a booster, SpaceX were much slower to do that than Blue Origin. This may have been because they had more at stake, with paying customers to think about rather than simple in-house test flights. While the first New Shepard was reused as early as January 2016, a Falcon 9 didn't follow suit until more than a year later, in March 2017. This was Falcon 9 flight 32 – and of the next 68 launches, almost two-thirds of them, 42 in total, reused boosters that had flown on at least one previous occasion. That's an amazing feat, which NASA hadn't even attempted with its own counterparts to Falcon 9, Atlas and Delta – or Roscosmos with Soyuz, or the European Space Agency with Ariane, for that matter.

As well as launching its own Starlink satellites, SpaceX has a number of outside customers for Falcon 9 launches. The highest profile of these is NASA itself, with flights of the Dragon resupply craft to the ISS occurring on a regular basis since 2012. Similar in design and appearance to the Apollo spacecraft that took astronauts to the Moon, the Dragon has now been upgraded to carry human passengers, and the first crewed flights to the ISS took place in 2020 – a subject we'll come back to in the next chapter.

But the Dragon and Starlink flights are just the best known of Falcon 9's activities. It's also launched batches of satellites for the Iridium company, GPS satellites for the US military and geosynchronous satellites for numerous customers around the world. The fact is, for any company with the resources to do it, launching hardware into orbit is big business.

SpaceX's first serious competitor was Orbital ATK,* which was contracted by NASA to provide 'commercial

---

* Now absorbed into Northrop Grumman, one of the giants of the US aerospace industry.

resupply services' to the ISS using its Antares launcher and Cygnus cargo ship alongside Falcon and Dragon. The idea of having two different companies providing essentially the same service was a novelty for NASA, but it makes good sense. Opponents of capitalism often disparage competition because they see it in purely monetary terms – as an incentive to cut costs, with all the negative connotations that entails. But competition can have a positive impact on quality. If a supplier knows that the customer always has two options, they'll do everything they can to make sure theirs is the better product. If they can't manage that, they'll try to excel in other ways, such as offering superlative customer service.

Another competitor SpaceX has to worry about is Blue Origin, run by Elon Musk's arch-rival Jeff Bezos. Although Blue Origin is starting out in a different market – suborbital tourism using New Shepard – we've already seen that its forward plans include a much larger rocket, New Glenn, to provide an orbital launch service. New Glenn is destined to be even more powerful than Falcon 9, with a payload capacity of 45 tonnes to LEO, but its first launch is still several years in the future.

The other player in the suborbital tourism market, Richard Branson, is also getting into the satellite launching business, but from a different angle.* Virgin Orbit's LauncherOne isn't trying to compete in the size stakes with Falcon 9 or New Glenn. It's much more modest in scale,

---

* Literally a different angle – horizontal instead of vertical.

comparable to SpaceX's original Falcon 1 test rocket, with a payload capacity of just 300 kg or so. Its novelty lies in the fact that it's air-launched, from a Boeing 747 transferred from one of Branson's other companies, Virgin Atlantic. In its first successful flight, on 17 January 2021, it put a batch of small educational satellites into orbit on behalf of the most venerable space customer of all, NASA.

The idea of launching satellites from an airborne platform isn't new. It was first used by Orbital ATK – or Northrop Grumman as it is now – with the Pegasus air-launched rocket. Pegasus has been in service since 1990, initially operated by NASA using a B-52 – the same type of aircraft that launched the X-15 rocket plane in the 1960s – and later by the company itself using a converted airliner. But during all that time, launches have been few and far between. In the whole decade from 2010 to 2019, there were just four of them, all on behalf of NASA itself. The fact is that the advantages of air launch, of which there are several, are probably outweighed by its disadvantages. Perhaps unsurprisingly, one of its most outspoken critics is SpaceX boss Elon Musk. In one discussion, after acknowledging that the speed and altitude of the plane do make the rocket's job easier, he went on:

> It's quite a small improvement. It's maybe a five per cent improvement in payload to orbit, and then you've got this humungous plane to deal with. Which is just like having a stage. From SpaceX's standpoint, would it make more sense to have a gigantic plane or to increase the size of the first stage by five per cent? I'll take option two.

**Northrop Grumman's Pegasus rocket, with
a NASA satellite on board, seen prior to
launch from its carrier aircraft in 2019.**
(NASA image)

Another problem comes when you try to scale things up for larger payloads. With a SpaceX-style rocket, you can simply make it bigger – as they did with the Falcon Heavy, which we'll come to in the next chapter. But there's a practical limit to the size of an aircraft, and they don't come much bigger than the ones that Northrop Grumman and Virgin Orbit are using at the moment.

Nevertheless, launching satellites from an aircraft does have some advantages over traditional rockets. For one thing, it becomes possible to launch from places where a Cape Canaveral-style launch complex would be impractical, such as 'Spaceport Cornwall' – near Newquay on the west coast of England – which is hoping to become a launch site for Virgin Orbit. At the same time there's an increasing market for lightweight satellites, thanks to the miniaturisation that's possible with modern technology. Air-launch aside, the 'small satellite' sector also has a place for smaller, less powerful rockets of the traditional kind – and this opens the door to start-up ventures that might not have billionaire entrepreneurs like Elon Musk or Jeff Bezos at the helm.

One such company is New Zealand-based Rocket Lab. Its Electron rocket is similar in specification to Virgin Orbit's LauncherOne, except that it's launched in the traditional way from the ground. In another respect, however, the rocket breaks a tradition that goes back to the German V-2 of the Second World War. That was the first successful liquid-fuelled rocket, at a time when scientists outside Germany firmly believed that such a thing was a physical impossibility. Their reasoning involved some complicated mathematics, but it essentially came down to the fact that to get the rocket off the ground, the propellant entering the combustion chamber had to be under such high pressure it would burst the fuel tank.

The devious trick the Germans used, which shocked the Allies when they learned of it, was to store the propellant at a much lower pressure, and then boost it as it was injected into the combustion chamber using a turbopump, analogous to the turbocharger on a car. Virtually every liquid-fuelled rocket since the V-2 has used a similar turbopump – but Rocket Lab's Electron is different. It uses an electric pump instead. This is a much simpler design mechanically, making it quicker and cheaper for the company to develop. Electron delivered its first payload to orbit in January 2018, and within two years had seen a dozen successful launches.

Over and above the number of flights achieved, Electron has been astonishingly prolific. Despite its relatively small payload mass, those first dozen launches put over 50 satellites into orbit, for multiple different customers. All those satellites had important work to do, but they weren't physically large or heavy. The same technology that allows us to pack a filing-cabinet-sized 1980s supercomputer into a 200-gram smartphone means that a satellite no longer has to be big in order to be useful.

Many of the satellites launched by Electron are 'cubesats' – literally made up of one or more standard cube-shaped units, just 10 cm on a side, and having a mass of 1.3 kg or less. Batches of cubesats, not necessarily all for the same customer, will also become the main payload for other small launchers, such as Virgin Orbit's LauncherOne. But small satellites don't have to be launched by a small rocket. SpaceX usually manages to squeeze a few cubesats in alongside the main payload on a Falcon 9 launch, for example.

One such cubesat has already been mentioned, in the box on page 59. That was the University of Surrey's RemoveDEBRIS demonstrator, which consisted of two separate satellites, each made up of two ten-centimetre cubes. It hitched a ride on a Falcon 9 that happened to be taking a Dragon resupply ship to the ISS in April 2018. The fact that cubesats are relatively cheap to construct and launch makes them particularly popular with cash-strapped academic research and educational establishments – and the democratisation of space doesn't end there.

How about an amateur satellite? You can't get more 'private enterprise' than that; it really is opening up the space sector for ordinary people. In one sense, amateur satellites have been around since the dawn of the space age. When a US government spy satellite was launched in December 1961, the rocket needed a small amount of ballast to balance it out properly. Instead of carrying a dummy weight, a group of engineers at the company that built the satellite had a better idea.

It just so happened that the engineers were enthusiastic 'radio hams', or amateur radio operators. In their spare time, they constructed a small, 10 kg satellite of their own to serve as ballast on the mission. Called Oscar 1, this only orbited for 20 days, and all it did was transmit 'hi' in Morse code over and over again. However, it did this in one of the frequency bands allocated for amateur use, so radio hams around the world heard the signal.

Later satellites in the Oscar series were more sophisticated, containing receivers as well as transmitters. This meant a ham in one country could send a signal up to the satellite and have it bounced back down to a counterpart in another country, who would have been beyond the range of a direct transmission. This might not sound very exciting in these days of Skype, Zoom and social media, but things were different a few decades ago. Amateur radio was the only way that groups of friends, or even complete strangers, in different parts of the world could have long, leisurely chats in real time – and it was a hugely popular hobby for much of the 20th century.

The amateur frequency bands are still around today, but their use has shifted more towards education rather than communication. Coupled with the affordability of cubesats, that's given a whole new slant to the idea of amateur satellites. For example, the FUNcube-1 satellite, just a single ten-centimetre cube, was designed and built by a group of radio amateurs in the UK and the Netherlands with the primary aim of supporting school-level education. As well as acting as an orbiting relay between amateur radio installations – which can now be found in many schools – the satellite also contains a small scientific experiment. Students

The FUNcube-1 satellite, built and operated by radio amateurs, is just 10 cm on a side.

(Wouter Weggelaar, CC-BY-3.0)

can download data from this, and compare with their own data from the same experiment conducted in the classroom.

FUNcube-1 was launched in November 2013, along with dozens of other cubesats, on board a Dnepr* rocket operated by a private Russian company, Kosmotras. This was the eighteenth of 22 Dnepr launches. With a total payload capacity of 4.5 tonnes to LEO, the three-stage Dnepr was a typical medium-sized space launcher, and would scarcely be worth mentioning except for one minor detail. The rockets were all built several decades before they left the ground, and they weren't designed to put satellites into orbit. They were designed to kill people – lots of people.

As mentioned in the first chapter, the world's first satellite, Sputnik 1, was launched into space on a repurposed intercontinental ballistic missile. Such missiles became the ultimate weapon of the Cold War, progressing through several generations before the East–West arms race eventually came to an end. The final generation on the Soviet side was called R-36M by its operators, and 'Satan' by the West. The latter name was entirely appropriate; the R-36M was capable of delivering eight megatons of death and destruction to any location on the planet. Thankfully it was never called on to do this, and when the missiles became surplus to requirements they were sold off (minus warheads) to Kosmotras. After a few tweaks to the control software – in the ultimate 'swords into ploughshares' transformation – Satan became Dnepr.

---

* This odd-sounding name comes from a river in eastern Europe, more commonly spelled Dnieper in English.

Thinking Outside the Box

Whether designed for military or peaceful purposes, rockets have become the standard way to get into space. The basic principle is very simple. You generate a huge amount of super-hot gas by burning fuel and oxygen, then eject it at high pressure through a small exhaust nozzle. By Newton's third law of motion ('for every action there is an equal and opposite reaction'), the momentum of the backward-travelling exhaust plume has to be balanced by forward momentum of the rocket itself. This works just as well in the vacuum of space as it does inside the Earth's atmosphere.

At the same time, there's something unsatisfactory about rocket travel. There's a nagging feeling that you're going about things in a wasteful and inefficient way. Why, for example, does the engine of a rocket plane like the X-15, or Virgin's SpaceShipTwo, run furiously for a minute or so and then run out of fuel, when an ordinary jet plane can fly for hours? And why do so many space launchers remain single-use, with even SpaceX's Falcon 9 being only partially reusable? Questions like these have led innovative thinkers to look for more efficient ways of getting into space – some further outside the box than others.

Probably the most way-out idea of all is the space elevator, which almost certainly falls in the 'interesting, but it's not going to happen' category. Recall that a satellite in geosynchronous orbit over the equator is rotating at exactly the same rate as the Earth itself. So what would happen if you dropped a long cable from such a satellite down to the surface of the Earth? In principle, you could then climb up the cable – at whatever speed you wanted, not some magic orbital velocity – all the way to the top. When you got there, lo and behold – you'd be in orbit.

Before you rush off to set up a company to do just that, it's only fair to point out a few snags. Firstly, it really does have to be a *very* long cable. The distance to GEO is around 36,000 km – not much short of the entire circumference of the Earth, which is 40,000 km. And to support its own weight without snapping, it would have to be stronger than any known material. Beyond that, it's not clear how much 'payload mass' the cable could support on the way up to orbit, or how fast the elevator cars would travel. You might be in for a long, slow ride.

Don't forget, too, that it's only when you reach the very top that you'll have enough centrifugal force to counteract gravity. When you're passing through the LEO altitudes, just 1 or 2 per cent of the way into your journey, there's nothing holding you up except the elevator cable. If that snaps, you're on your way back down again, Felix Baumgartner style.

As impractical as space elevators may seem – here on Earth, anyway* – it's a concept that has been looked into by a number of groups around the world. This isn't because they have serious hopes of building one in the foreseeable future, but to try and crack some of the most obvious

---

* The physical challenges of a space elevator would be more tractable on the Moon or Mars, but in those cases there's the added complication of undertaking a major engineering project at such a remote location.

technological challenges – such as high-tensile strength fibres, or cable-climbing robots – which might have other, more immediate applications.

In this vein, a tiny space elevator was even launched into Earth orbit in September 2018. It's called STARS-Me, which stands for (take a deep breath) 'Space Tethered Autonomous Robotic Satellite – Miniature Elevator'. The work of Shizuoka University in Japan, it consists of two single-unit cubesats separated by a 14-metre tether. The 'elevator' takes the form of a small robotic crawler which shuttles back and forth along the tether.

'Thinking outside the box' doesn't have to mean looking for some super-futuristic concept like a space elevator, though. What about a much older technology, in the form of balloons? We've already seen how rockets like Pegasus or LauncherOne can be launched into space from jet aircraft, so how about launching them from helium balloons? These can reach higher altitudes than most jets – well into the stratosphere, where the air is thinner and the rocket's job is easier – and they use up far less energy in getting there.

In 2019, a Welsh start-up company named B2space announced its intention to offer just such a service. Their balloon would ascend to a height of 35–40 km, and then launch a rocket to carry a small, cubesat-type satellite into orbit. By launching from such a high altitude, the company estimates the rocket would use 85 per cent less fuel than it would if it took off from ground level.

The downside, of course, is that the balloon itself has a limited lifting capacity, so the system would only be practical for launching cubesats or similarly small payloads. There's no way it could ever be scaled up to carry a crewed spacecraft, or a large geosynchronous satellite. For those, we're still stuck with multi-stage, ground-launched rockets. Or are we?

The idea of a 'single-stage to orbit' (SSTO) vehicle has been a dream of aerospace designers for decades. It's such a simple idea you feel it ought to be possible. When an airliner lands at its destination, it looks exactly the same as it did when it took off. The only difference is that there's less fuel in its tanks. As soon as it's refuelled, it can take off again. But this just isn't possible with space launchers – not outside sci-fi movies, anyway. At first sight NASA's Space Shuttle appeared to come close to the SSTO goal, but that was just smoke and mirrors. When it took off it had a huge external fuel tank, and two solid rocket boosters, that weren't there when it came back. Sorry, but that doesn't count as 'single-stage'.

Over the years, various government-funded projects have tried and failed to come up with a solution to the SSTO problem. Various private-sector companies have tried, too – and one of them believes it's cracked it. This is another British firm, Oxfordshire-based Reaction Engines Limited, which was founded in 1989 and has worked on the SSTO problem ever since. Its solution is a spaceplane called Skylon, which is visually a little like the Space Shuttle – except that it doesn't need disposable boosters or an external fuel tank. The idea is that it would take off from a runway like a conventional aircraft, fly all the way into orbit to deploy its payload, then return to a landing on the same runway. In other words, it would accomplish the standard SSTO dream just as it's depicted in sci-fi movies.

The trick is that, when Skylon is doing something that a conventional jet could do perfectly well – like ascending through the dense lower parts of the atmosphere, or descending back down again – it actually *is* just a conventional jet. It climbs by exploiting aerodynamic lift on its wings, and its engine draws in large amounts of air from the atmosphere, to provide both the necessary exhaust mass and the oxygen needed to burn the spaceplane's liquid hydrogen fuel.

Then at around 30 km altitude, where the air becomes too thin for a jet to work efficiently, Skylon's engine switches into rocket mode. Instead of trying to breathe the increasingly thin air, it switches to its onboard oxygen supply to run the engine – which, incidentally, is called SABRE, for 'Synergetic Air-Breathing Rocket Engine'. Actually, that's a contradiction in terms, because a rocket engine isn't a rocket engine if it breathes air. By definition, that turns it into a jet engine. A little theory may help to clarify things at this point.

You already know one of the reasons why air is so important to an engine. As you've no doubt noticed, your car gets a lot livelier when you stamp your foot on the accelerator. All this is doing is increasing the air flow into the combustion chamber. A jet engine uses air in much the same way, but that's just one of the reasons it needs as much air as it can get. Only a fraction of the air it draws in actually goes into the combustion chamber; most of it remains at ambient temperature, and gets blasted out at high speed through the exhaust fan (which is turned by the heated gas from the combustion process, in case you're wondering). In rocket jargon, the air blown out through the fan is 'propellant' – but no one in the jet business ever bothers to call it that, because they get it for free.

This explains why SABRE is so much more efficient than a conventional rocket. Whenever it can, it gets as much of its propellant as possible from the outside air, only resorting to its onboard supply when it absolutely has to. That's completely different from traditional rockets, which carry everything along with them right from the start. It's why they need multiple stages, or external fuel tanks and expendable boosters.

According to Reaction Engines' estimates, Skylon would be capable of delivering a 17-tonne payload to LEO, much like SpaceX's Falcon 9, but for around two-thirds of the cost. And in contrast to Falcon 9, of which only the booster stage can be reused, Skylon would be a completely reusable vehicle, with a turnaround time between flights of a few days. All in all, Skylon looks like the perfect solution to everyone's space launch needs – except for one tiny detail. It doesn't exist yet.

Of course, there are dozens of brilliant ideas that don't exist yet (remember space elevators?) – but in the case of Skylon there's more cause for optimism. Reaction Engines is a well-established firm that, as well as pursuing the Skylon idea for three decades, has conducted enough other business to keep its head above water. In the three years 2017–2020, the company raised over £100 million from various sources, allowing it to progress from pure research on the SABRE technology to development and testing of core engine components. Plans for a flying test bed for the SABRE engine are also well advanced.

Unfortunately, with the relatively slow rate of investment available to Reaction Engines, it will still be many years – maybe decades – before Skylon makes it into orbit. By that time, it might have missed the boat. As inefficient as traditional rockets are, they could still end up being the cheapest way to put hardware into space, simply through economies of scale. If components are mass-produced, and reused wherever possible, then as soon as research and development costs have been amortised, the cost of sending a payload into orbit could drop dramatically.

That's the philosophy of one person in particular – Elon Musk of SpaceX. He has pretty big plans for the future, as we'll discover in the next chapter.

VACATIONS IN SPACE 4

Astronaut crews have been flying up for extended tours of duty on the International Space Station ever since Expedition 1 in November 2000. But 20 years later, the flight that took off from Cape Canaveral on 16 November 2020 was different. It didn't use Russia's state-operated Soyuz work-horse, or America's state-operated Shuttle, which had been retired back in 2011. It used the world's first privately built spacecraft to be rated for human-carrying flight, SpaceX's Crew Dragon. The astronauts themselves may have been government employees – Mike Hopkins, Victor Glover and Shannon Walker from NASA, and Soichi Noguchi of the Japanese space agency – but the technicians at the launch-pad, and the people behind the desks at mission control, all worked for SpaceX.

Launched atop another SpaceX product, the tried-and-tested Falcon 9, *Resilience* was the second Dragon capsule to carry a crew into space. It followed a short test run, using

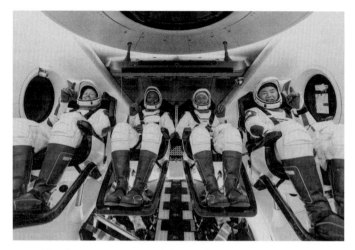

**The four astronauts of SpaceX's Crew-1 mission,
inside their Dragon capsule, *Resilience*.**
(NASA image)

Crew Dragon *Endeavour*, which carried two astronauts to
the ISS and back six months earlier. Prior to that, uncrewed
Dragon capsules had flown to the ISS on numerous occasions, as part of the commercial resupply service that NASA
had contracted to SpaceX, alongside another private firm,
Orbital ATK. The new crew transfer service follows similar
lines, with NASA once again hedging its bets by placing similar contracts with two independent suppliers.

In this case, the second supplier is Boeing, the largest
aerospace company in the world. With a history going back
to the First World War, it got into space manufacturing right
at the start, with a contract from NASA to construct the first
stage of its giant Saturn V rocket. As we saw earlier – and in

common with the lucrative defence contracts that pulled in money for Boeing and the other aerospace giants of the Cold War era – this was done on a no-risk cost-plus-fee basis. The resulting working culture may have had advantages, but completing projects on schedule and to budget weren't among them.

Contracting practices may have changed, but ingrained traditions die hard. While SpaceX gets things done quickly, Boeing takes its time. Ever since 2011, it's been endeavouring to build a modern-day successor to the Saturn V called SLS (for Space Launch System). This still hadn't got off the ground ten years later, during which time SpaceX had progressed from a rookie on the space launch scene to its most reliable player.

Boeing's counterpart to Crew Dragon, a similar-spec capsule which it calls Starliner, is progressing a little faster than SLS, although it's at least a year behind its SpaceX competitor. Starliner's first, uncrewed, test flight took place in December 2019, but it was a partial success at best. It managed to get into orbit, but it was the wrong orbit to complete its intended mission, which should have included a visit to the ISS.

Undoubtedly Boeing will get things done in due course, to meet its contractual obligations to NASA. But the culture at SpaceX, under its charismatic if unconventional leader Elon Musk, is completely different. Crew Dragon isn't just a pragmatic way to fulfil a government contract. It's a privately operated, human-rated spacecraft that SpaceX can utilise in a range of ways – of which meeting the NASA contract is just one. Some of the others are a lot more exciting.

The Return of the Space Tourist

Crew Dragon can carry up to seven people at a time, but the NASA-sponsored flights ferrying astronauts to the ISS will usually leave several seats empty. That's a commercial opportunity in itself. Remember what happened on Soyuz flights, in the days when it was just a second string to the Shuttle? Quite often there was a free seat, which was rented out to paying passengers like Dennis Tito and Charles Simonyi via Eric Anderson's company Space Adventures, Inc. That opportunity dried up when the Shuttle was retired, because after that every seat on a Soyuz flight was needed for ISS astronauts. Now, with Crew Dragon on the scene, the day of the orbital space tourist is with us again.

Like SpaceX's Falcon 9, Crew Dragon is reusable. The *Endeavour* capsule, used on the first test flight, returned to the ISS with another batch of astronauts in April 2021. But the second flight of *Resilience*, scheduled for September 2021, won't have any professional astronauts on board at all. Its crew will consist of billionaire Jared Isaacman, who is footing the bill, and three other private citizens. The mission is dubbed Inspiration 4 and aims to be the first purely sightseeing mission in orbit. And that's just the start.

It seems likely that Space Adventures will revive their old business model, offering seats on Soyuz missions that have been freed up by the availability of Crew Dragon, and/ or some of the spare seats on Dragon itself. But the opportunities don't stop there. Roscosmos, the Soyuz operator, has only ever been a grudgingly reluctant provider of space tourism services, driven by the lure of a little extra money rather than an enthusiastic desire to open up an exciting new market. With SpaceX and Elon Musk, the situation couldn't be more different. For them, Crew Dragon offers far more opportunities than simply selling spare seat capacity on government-sponsored flights.

There's already an agreement between SpaceX and Space Adventures to fly a tourist-only mission, carrying four passengers into orbit for three or four days. The price per passenger is likely to be comparable to that paid by previous orbital tourists – an eight-figure sum in US dollars – but the amount of pre-flight training they will have to undertake is a lot less, due to the greater comfort and simplicity of Dragon as compared to Soyuz.

The flight will be novel in a couple of other ways, too. For one thing, the four passengers will be the only people on board. That may sound alarming ('Don't look at me – I thought you were the pilot'), but remember that Dragon spent a decade flying itself before it ever carried a human crew. The other novelty, and a much more exciting one, is that if everything goes according to plan the passengers will orbit the Earth at a higher altitude than any humans before them. The current record is 1,400 km, set by NASA's Gemini 11 crew in 1966. The intention is that the Space Adventures Crew Dragon flight will go higher than that. Obviously, the astronauts who went to the Moon were further away than this, but they weren't in Earth orbit – so it will still be a genuine first in that respect.

No firm date has been set for the flight, not least because the passengers haven't been selected yet, but it's been pencilled in for 2022 at the latest. In the same time frame, it's possible that Crew Dragon may take its first paying passengers to the ISS as well – and in this case, they won't simply be tourists on a sightseeing trip. Organised not by Space Adventures but by another company called Axiom Space, the primary purpose of the flight is to film the first ever movie scenes in space. The film in question – the existence of which was revealed in May 2020, although its title wasn't – will star Hollywood veteran Tom Cruise. He'll be one of the passengers going on the Axiom/SpaceX flight, along with director Doug Liman.

The fact that SpaceX is happy to fly Hollywood movie-makers into space, for an appropriate fee, comes as no surprise. What really is surprising is that the operators of the ISS – which, after all, is a highly specialised working environment for scientific research – are prepared to have the likes of Tom Cruise bouncing around in their midst filming an action movie for several days. But it seems they think the potential benefits will outweigh any disruption the activity might cause. As NASA administrator Jim Bridenstine said on Twitter:

> NASA is excited to work with Tom Cruise on a film aboard the Space Station. We need popular media to inspire a new generation of engineers and scientists to make NASA's ambitious plans a reality.*

---

* And NASA isn't the only space agency with an eye on movie stardom. As the final touches were being put on this book, Roscosmos

A few months after the Tom Cruise announcement, Axiom revealed another project in the offing – albeit one that is still several years in the future, and might or might not involve SpaceX. It's possible they may use Boeing's Starliner instead for this one. Here again, the client isn't a space tourist in the usual sense, but a TV company planning to pick a member of the public via a reality show and send them on a trip to the ISS as a prize. The show will be titled *Space Hero*, and the production company, specially set up for the purpose, calls itself Space Hero, Inc.

Whether anything will come of this remains to be seen. At first glance, it looks a little too close to previous projects that have promised far more than they were able to deliver. First, in December 2005, came a British TV show called *Space Cadets* – produced by the same company, Endemol, that had been responsible for the grandaddy of all reality shows, *Big Brother*. The basic concept was ostensibly identical to *Space Hero*. A group of contestants, confined to a sealed-off environment in what appeared to be a private space centre, were whittled down until the final four were selected for a five-day trip into Earth orbit.

That's what the participants were led to believe, anyway. Right from the start, the audience at home were let in on the secret. The whole thing was a hoax, and the spaceflight would be simulated on Earth using a plywood space shuttle that had previously seen service as a prop in films like *Armageddon* and *Space Cowboys*. The contestants were put

announced a similar venture involving the Russian actress Yulia Peresild and director Klim Shipenko.

through a series of tests which, unbeknownst to them, were designed to select the exact opposite of the character traits that would really make good astronauts.

The producers wanted people who were, to put it bluntly, stupid enough that they wouldn't question why they didn't feel any g-forces during launch, why they didn't become weightless in space, or why they weren't allowed to look through the windows used by the shuttle's 'crew' (actually played by actors). The tests were designed to select candidates who not only lacked a scientific education, but had no trace of scientific curiosity or judgment either. On top of that, they had to be sufficiently suggestible to accept anything they were told, no matter how ludicrous, rather than trusting the evidence of their own senses.

Even so, many viewers found it hard to believe that anyone could be quite as gullible as the winning contestants appeared to be. Unlike most reality shows, there was no audience involvement – no voting, no calls to premium phone lines – and hence no cause for litigation on grounds of misrepresentation. So it seems likely that *Space Cadets* was actually a reverse hoax – not on the gullibility of the participants, who were probably in on the joke all along, but of the TV audience who lapped it up.

Whatever kind of hoax *Space Cadets* was, it was ultimately a credible one. The experience on offer, a flight into Earth orbit on board a space shuttle, was perfectly possible with tried and tested technology that existed when it was aired in 2005. The same can't be said of a later crossover between the worlds of reality TV and space travel: the notorious Mars One initiative. Here the aim wasn't simply to send contestants into space, but all the way to the Red Planet.

The concept, announced to much fanfare in 2012, was the brainchild of Bas Lansdorp – who, like reality TV pioneers Endemol, hailed from the Netherlands. Lansdorp wasn't a space scientist but a businessman, and he treated the voyage to Mars as a business problem rather than a scientific one. His actual understanding of space travel seems to have been quite sketchy – possibly derived from watching sci-fi movies, where travelling to other planets is a doddle once you're 'in space'. In the real world, even getting to the Moon is hard enough, which is why so few people have been there. Getting to Mars is harder still, and no one has ever managed that.

Even from a business point of view, Lansdorp's reasoning looks shaky – confusing the profitability of a reality TV series, which might be measured in hundreds of millions of dollars over several years, with the total cost of going to Mars, which could be hundreds of billions. That's a thousand times as much. Add the fact that Lansdorp put more effort into headline-grabbing activities like the selection of candidates, rather than working out how he would actually get them to Mars, and the outcome was all but inevitable. Mars One was declared bankrupt in January 2019, with a total debt of approximately €1 million.

The sad thing about Mars One is that, buried somewhere in all of Lansdorp's muddled thinking, there's actually a brilliant idea. There's nothing fundamentally wrong with setting a reality TV show in space. It would be simultaneously entertaining and educational, raising awareness and enthusiasm about the possibilities of space among the general public – and probably earning a decent profit in the process.

Where Mars One slipped up was in trying to do something – send humans to another planet – that even well-established space agencies had never succeeded in doing. It was essentially an impossible undertaking, certainly in the projected time frame. If Lansdorp had stuck to some space-based activity that's been done, safely and reliably, on a routine basis by many people in the past – such as visiting the ISS – his idea might have worked out. So it's possible that Axiom's *Space Hero* will end up faring better than Mars One after all.

'But the ISS is boring,' you might be thinking. 'I want to go to the Moon. You can't say that's impossible, because Neil Armstrong went there more than 50 years ago.' It's funny you should mention that, because other people have had exactly the same thought – as we'll see now.

Long-haul Flights

For a given payload, the further away from Earth you want to send it, the bigger the rocket needed to launch it. That, in a nutshell, is the essence of rocket science. Whoever first used the phrase to mean 'too difficult for a non-expert to comprehend' really ought to be shot, because it's not true at all. Compared to other branches of applied physics, such as electronics or aerodynamics, rocket science is simplicity itself.

It all comes down to the energy budget. As we saw in the first chapter, getting into Earth orbit means gaining both height (potential energy) and speed (kinetic energy). The higher the orbit, the more energy you need in order to reach it. There's only one place this energy can come from, and that's the fuel in your rocket. In principle it would be possible to refuel the rocket in low orbit, allowing it to boost itself up to higher altitude, but this is unlikely to become standard practice in the foreseeable future. At the moment, for a given size of rocket, you're stuck with the fact that the size of payload you can put into a high orbit, such as GEO, is going to be significantly smaller than the payload you can put into a low orbit. If you want to do better than that, you're going to have to use a bigger rocket.

The same holds true when you look at longer journeys, beyond Earth orbit. Or it does if you want to get to your destination as quickly as possible, which will usually be the case with a human crew. It's possible to reduce the energy needed for an interplanetary trip by taking a roundabout route, gradually spiralling towards your destination over the course of years, but that's only a viable option for robot probes. They can be switched off during the journey, so they really don't care how long it takes. For humans, on the other hand, the basic equation still holds: the further away the destination, the more rocket fuel is needed. It takes a bigger rocket to get to the Moon than to LEO or GEO, and a bigger rocket to get to Mars than to the Moon.

The problem for space tourists hoping to travel beyond Earth orbit is that the rocket-building business, like any other, is driven by the law of supply and demand. Manufacturers only make launch vehicles if there's a sufficiently lucrative market for them – and there just isn't that much call for the kind of super-heavy launcher that can take a human crew to the Moon or beyond. NASA's Saturn V was built specifically for the Apollo missions, but it had little practical utility beyond that. Its payload to LEO was a record-breaking 140 tonnes, but how many customers need to put 140-tonne satellites into orbit? In the end, the Saturn V was only ever used once after the final Apollo mission, to launch NASA's massive Skylab space station in 1973.

Until recent years, when people started thinking seriously about space tourism, there was only one acceptable rationale for interplanetary travel: scientific research. This doesn't call for particularly powerful rockets, because – in the eyes of NASA and the other national space agencies – planetary exploration is the work of robots, not human beings. This means the payloads involved are actually quite small. NASA's largest Mars rovers, Curiosity and Perseverance, were launched by Atlas V rockets, comparable in size to SpaceX's Falcon 9. The reason is that the rovers only weigh a tonne or so each, compared to 45 tonnes for the Apollo spacecraft – or considerably more than that for a long-duration spacecraft capable of carrying a crew all the way to Mars.

Such flights are going to require a super-heavy launch vehicle, and when it comes down to it, they're the strongest justification for building such a thing. Once you've got one, there are other things you can do with it – such as launching a dozen large geostationary satellites in one go – but those are all things you could have done with multiple launches of a smaller rocket. To go to all the effort of designing and constructing something larger, you need an aerospace manufacturer that actually sees crewed interplanetary flights as a high priority.

Enter Elon Musk and SpaceX. Since 2016, the company's motto has been 'Making Life Multiplanetary', and Musk is serious about it. The first step in building larger rockets was the Falcon Heavy, essentially three Falcon 9 boosters strapped together. Not surprisingly, its capacity to LEO is roughly three times the Falcon 9's, or about 64 tonnes. This may be less than half the Saturn V's limit, but it's more than twice that of any other rocket currently in service.

As simple as it sounds in principle, actually building the Falcon Heavy and getting it off the ground was quite a challenge. In an interview prior to the first test flight, Musk put it this way:

> It actually ended up being way harder to do Falcon Heavy than we thought. At first it sounds real easy, you just stick two first stages on as strap-on boosters. How hard can that be? But then everything changes. The loads change, aerodynamics totally change … really way, way more difficult than we originally thought.

Even at the time of that maiden flight, SpaceX weren't confident the Falcon Heavy would make it into orbit. In the same interview, Musk merely said 'I hope it makes it far enough away from the pad that it doesn't cause pad damage.

I would consider even that a win, to be honest.' This put the company in something of a quandary. They needed to test the new rocket, to demonstrate both to themselves and to potential customers that it worked. But then again, it might not work – and if it didn't, no one wanted to destroy a valuable payload along with the rocket.

It's a situation that often crops up with the first flight of a brand new rocket, and the usual solution is to launch a dummy payload of the required mass. That's what happened with SpaceX's very first launch, of the diminutive Falcon 1 rocket in 2009. It simply put 165 kg of dead weight into orbit. To do the same with Falcon Heavy, on the other hand, would involve putting 64 tonnes of dead weight into LEO – or 26 tonnes into GEO – and that would be downright antisocial. It would be that much more space junk cluttering up orbits that are needed by working satellites.

So Musk came up with a better idea. After all, Falcon Heavy wasn't designed to be just another satellite launcher – it was the first step in 'making life multiplanetary'. So if it had to launch a dead weight into space, why not put it on an interplanetary trajectory? And why not make that 'dead weight' something a little more memorable than a conveniently shaped hunk of metal? Although Musk has only featured in this book so far in the context of SpaceX, he's better known to the world at large as CEO of the Tesla electric car company. So why not take the opportunity to plug Tesla at the same time as demonstrating Falcon Heavy's capabilities?

At the time of that first flight in February 2018, Musk's own Tesla Roadster was ten years old and beginning to show its age. Most people know what happened next. The car ended its career in spectacular style, as the 1,300 kg dummy payload on Falcon Heavy's maiden flight – which, despite all the reservations, was a complete success. The car was a featherweight payload for a rocket of that size, which in consequence was able to launch it onto a truly enormous orbit.

The rocket easily gave the car enough energy to escape the Earth's gravity, so it ended up on an orbit around the Sun rather than the Earth. Musk's original intention was that the car would loop symbolically between the orbits of Earth and Mars, but in the event the rocket gave it even more energy than intended. The actual outer extremity of the trajectory – which the Roadster reached in November 2018, nine months after launch – is ten per cent further away than Mars.

As fake as this picture looks, it's a genuine photograph of Elon Musk's Tesla Roadster, complete with spacesuited mannequin, as it headed away from Earth after the maiden flight of Falcon Heavy.

(SpaceX image)

Even before Falcon Heavy got off the ground, Elon Musk was busy planning the new rocket's first space tourism flight. In February 2017, he announced that SpaceX would send a then-unnamed passenger on a sightseeing trip around the Moon before the end of the following year, 2018. That was a pretty tight schedule, given that not only had Falcon Heavy not flown yet, but neither had Crew Dragon – the other essential component of the proposed mission.

This wasn't the first time such a bold announcement had been made. Something similar happened in 2012, when a company called Excalibur Almaz declared its ambition to send passengers on trips around the Moon within a few years. Although the company's owners were predominantly American, it was actually based on the Isle of Man. For those who might not know, that's a small island halfway between Britain and Ireland, better known as a holiday destination than a hotbed of cutting-edge aerospace research. Nevertheless, as Excalibur Almaz CEO Art Dula explained, 'We got a very nice deal on rent from the Isle of Man government.'

There was a reason Dula and his colleagues believed they could get their business up and running in no time flat. They already had most of the hardware they needed, and it had already been thoroughly space-tested. In effect, they were playing another swords-into-ploughshares game like the Dnepr launch vehicles mentioned in the previous chapter. Those rockets started life as part of the Soviet Union's arsenal of ballistic missiles, and Excalibur Almaz had likewise bought up a hangarful of ex-Soviet military stock.

In this case, it all came from a formerly top secret project called Almaz (hence the company's name). Developed by the Soviet military in the 1970s, this was intended to be a series of orbiting space stations, together with a fleet of reusable spacecraft to ferry crew and equipment to them. In the end, however, only two stations ever became fully operational, since the Soviet government decided it could achieve the same ends, at lower cost, using spy satellites. By this time a fair-sized stockpile of now-redundant hardware had been built up, in the form of spacecraft and station modules – and these were what eventually ended up in the hands of Excalibur Almaz.

Unfortunately, the whole enterprise collapsed within a few years. Or make that 'fortunately', if you don't fancy the idea of flying into space in a forty-year-old relic of the Cold War. That seems to have been how Excalibur Almaz's potential customer base felt, anyway. The company never did get any bookings, and by March 2015 it had been forced to sell off all those second-hand spacecraft. The buyer remained anonymous, but let's hope it was a museum this time.

Not surprisingly, SpaceX's lunar plans have fared some-what better. Although the initial 'end of 2018' time frame was over-ambitious, they could probably have made the end of 2020 if they'd gone all out for it. By that time Crew Dragon was tried and tested – it had flown into space twice with people on board – as was Falcon Heavy. As well as that memorable first flight, it was launched twice in 2019 with satellite payloads for paying customers: first a 6.5-tonne geostationary satellite for Saudi Arabia, followed by a cluster of satellites for the US air force.

The proposed trajectory around the Moon would also present no difficulties using current SpaceX technology. When we hear that phrase 'around the Moon', it's tempting to think that it means going into orbit around the Moon, which isn't a trivial thing to do. When the spacecraft reaches the Moon it has to fire its rocket engines to slow itself down, in just the right way that it enters the desired orbit. Then, when all the sightseeing and selfie-taking is over, the engines have to be fired again in order to set the course back to Earth – another crucial manoeuvre that, needless to say, has to be done exactly right.

This is what NASA's Apollo 8 mission did in December 1968, seven months before the first lunar landing. While the Apollo capsule itself was somewhat smaller than Dragon, it had a much more powerful rocket motor for entering and leaving lunar orbit. So Dragon wouldn't be able to do that. But it can still go 'around the Moon' without entering orbit, if it's launched on a so-called free-return trajectory. This uses the Moon's gravity to swing the spacecraft back towards Earth with only a minimal expenditure of rocket power. The Apollo missions were all designed with such trajectories as a backup option in case of emergency – and one of them, Apollo 13, actually needed it, after its main engine was damaged en route. All in all, a free-return trajectory is the safest way to get to the Moon and back, and it's the route SpaceX's tourist would take.

That flight is still on, but it's been delayed by several

years because SpaceX's plans have changed. To put it bluntly, they've become positively grandiose.

From Dragons to Starships

SpaceX made the original announcement about the lunar tourist trip using Falcon Heavy a year before that rocket's first flight. By the time the flight took place, Elon Musk had moved the goalposts. He decided that, rather than seeking to have Falcon Heavy certified for human flight, it would only be used for satellite launches. As a result, Crew Dragon would be limited to Falcon 9 launches, and hence to Earth-orbital missions. For longer trips with human passengers, Musk now had bigger plans – much bigger ones.

From the early days of SpaceX, even before the first launch of Falcon 1, Musk had talked about building a gigantic rocket that would dwarf NASA's iconic Saturn V. Over the course of time this was referred to by a variety of names, including Big F——ing Rocket, Big Falcon Rocket, Mars Colonial Transporter and Interplanetary Transport System. Most people assumed it was just a pipe dream, or at best an aspirational goal for some indeterminate time in the future. It was only when SpaceX actually began hardware testing in 2019 that it became clear they were serious about the project. It even acquired a definitive name: Starship.

SpaceX's Starship will stand out from the crowd in several respects. As well as being enormous, more than 120 metres in height, it will be the first launcher since the

Saturn V to be designed primarily for crew-carrying missions beyond Earth orbit. It's also unusual in that it combines both launcher and spacecraft – Saturn and Apollo, if you like – in a single unified design. In this respect, it's reminiscent of a larger, more thoroughly reusable, version of the Space Shuttle.

If you remember how the Shuttle worked, it was made up of a three-engined orbiter plus two solid rocket boosters. At launch, both the boosters and the main engines had to fire up simultaneously to get the Shuttle off the ground. A couple of minutes later, the boosters dropped away to leave the orbiter to fly up into orbit, dropping off its external fuel tank on the way. It was a messy system, and it wasn't the original plan. The first Shuttle design involved just two components, both fully reusable, to be employed sequentially. First there was a winged booster to get the combined vehicle off the launchpad, with the booster then detaching and flying back to a horizontal landing when its fuel was used up. After the booster separated, the orbiter's engines would fire up to take it the rest of the way to orbit.

It's a more elegant solution than the Shuttle that was eventually built, but it was much too complex to be practical – consisting, in effect, of two large vehicles that had to function both as vertically-launched rockets and as horizontally flying aircraft. But that was back in the days when the only reliable way to land a space launcher was to fly it down onto a runway. Now that SpaceX's Falcon rockets routinely land in an exact reversal of the way they take off – vertically under rocket power – it's possible to revisit the original 'two-stage, fully reusable' shuttle concept and make it practical and affordable. That, in essence, is what SpaceX's Starship is.

The Starship proper is the upper part of the ensemble, which combines the payload-carrying spacecraft and the launcher's second stage in a single 50-metre unit – half the length of a football field. The first stage, called Super Heavy, adds another 72 metres to the height, making 122 metres in all – slightly taller than the Apollo–Saturn V stack, which set the benchmark for giant rockets back in the 1960s. Like the first half of the original shuttle design, the Super Heavy will get Starship off the ground, before separating and returning to base. Then Starship's own engines will fire up to boost it into Earth orbit – and they can be used again to take it further if required.

Both Super Heavy and Starship use the same type of rocket engine, specially developed by SpaceX for this task. Called Raptor, there will be a cluster of 28 of these powering Super Heavy, and another six on Starship itself. The most notable thing about the Raptor engine is its fuel, which isn't one of the two standard fuels, liquid hydrogen and kerosene, normally used in large rockets. Instead, Raptor burns methane, which SpaceX consider to be the most cost-effective fuel for a reusable vehicle. Methane has a second benefit, too, related to one of Musk's principal reasons for building Starship in the first place: his dream of flying humans to Mars. It just so happens that methane is the easiest fuel to synthesise on Mars, from raw materials that are readily available there.

As with Dragon, there will be both crewed and uncrewed versions of Starship. Both will be needed for really long trips, such as that journey to the Red Planet. In this case, the first flights to Mars – several of them – would be uncrewed robotic missions, aimed at setting up a propellant plant to synthesise plenty of methane and oxygen from the materials available in situ. Then two more Starships would take off from Earth, one carrying the crew and passengers, the other a spare load of fuel and oxidiser. The latter would be transferred to the crewed Starship in Earth orbit, giving it everything it needed for the onward journey to Mars.

Because Starship knows how to land vertically under its own rocket power, getting it down onto the surface of Mars is no problem. Once there, it can refuel using the previously prepared propellant plant, and then – thanks to the lower surface gravity of Mars – take off again without needing an additional booster, and head back to Earth.

The whole thing is much simpler than previously envisaged flights to Mars, which generally employed three different vehicles for transit, landing and ascent from the Martian surface. Starship performs all three functions in one package. The crewed version has a huge interior volume, over 800 cubic metres, so the trip out to Mars and back will be more like a pleasure cruise than the claustrophobic experience of traditional astronauts. According to SpaceX's own blurb, 'the crew configuration of Starship includes private cabins, large common areas, centralised storage, solar storm shelters and a viewing gallery.'

SpaceX's planned Starship will be large enough for passengers to enjoy weightless entertainment, as seen in this artist's rendering.
(SpaceX image)

As inspiring as the Mars plans are, they still lie quite a way in the future. In its early days, Starship will have to earn its living in more mundane ways, such as putting huge payloads into orbit. With a capacity of 100 tonnes or more, this might comprise, for example, one or two geostationary satellites, a sizeable constellation of LEO satellites and hundreds of cubesats – all for different customers. The deployment process would be more like that of the Space Shuttle than a conventional rocket, with the nose of Starship opening up like a clamshell to reveal a huge, thousand-cubic-metre payload bay.

So what about that flight of a paying tourist around the Moon, originally scheduled for Crew Dragon and Falcon

Heavy? It's still on, but with Starship in the offing it's been transformed into a much more ambitious project. The person footing the bill, reputed to be so enormous that it's made a significant contribution to Starship's development costs, was eventually revealed to be Japanese billionaire Yusaku Maezawa. His fortune has nothing to do with the world of science or technology, having been made in the fashion business, and Maezawa is more interested in the arts than the sciences. It's from this perspective that he's approaching his hoped-for trip around the Moon.

In the original plan involving Crew Dragon, Maezawa would have taken just one or two other people with him. But all that cabin space on Starship opens up bigger possibilities. After all, if Maezawa is paying for the trip, he might as well take along as many fellow passengers as he can. His plan now is to take up to eight other people with him – but not just any people. 'I choose to go to the Moon with artists,' Maezawa announced in September 2018. His hope is that this motley crew of bohemians will be inspired to create great new works of art by their circumlunar experience – a project Maezawa now refers to by the hashtag #dearMoon.

Lasting six days, the #dearMoon flight – like the Space Adventures Crew Dragon mission to 1,500 km altitude mentioned earlier – will be the extraterrestrial equivalent of an ocean cruise. That's all very well, but many vacationers prefer the idea of travelling to an actual destination and staying in a hotel. What are the prospects for an outer space counterpart to that experience?

Hotels in Space

This book opened with Arthur C. Clarke's futuristic vision of a luxury hotel in Earth orbit. Will such a thing ever be possible in the real world? It's an idea that many companies have looked into over the years. Interestingly, these aren't the same companies, such as SpaceX or Virgin Galactic, that have gone into the space launch or spacecraft-making business, but ones that have focused specifically on the space hotel angle. The fact is that the necessary specialisms are somewhat different – more closely related, in some ways, to earthbound civil engineering than to the established aerospace industry.

So what's actually required? The essential thing is a sufficiently large pressurised volume to contain the living space and usual facilities for the hotel guests. On top of that, because this particular hotel is in outer space, it needs life support and environmental control equipment, plus a way for visitors to transfer between the hotel and the spacecraft they arrive and depart in – and, of course, a safe way to evacuate in case of emergencies.

As it happens, all these things already exist in tried and tested form on the ISS. This was built as a working environment, of course, so it's more akin to a suite of offices or research labs than a hotel. But it does show that such a thing is technically possible. The problem is, it shows something else too – that it's not a cheap or easy thing to put together. It took more than twenty Shuttle flights, spread out over a decade, to get even close to its final form. The total cost of the station, shared between the participating nations, was around $80 billion.

Fortunately, a space hotel doesn't have to be anything like as expensive or time-consuming to construct as the ISS. One reason is that it's so much simpler; it doesn't need any of the scientific equipment that makes the ISS so complex and costly. Another reason is that, by using technologies that didn't exist when the ISS was designed, it can be made in a way that's much simpler and quicker to deploy into orbit.

Of the various companies eyeing the space hotel market, the one most likely to succeed is Bigelow Aerospace. Despite the name, its founder, the billionaire Robert Bigelow, doesn't have a background in space or aeronautics but in the tourism industry. In the present context, that's not a bad thing at all. Most of Bigelow's money came from another business he owns, the hotel chain Budget Suites of America.

Since the early 2000s, Bigelow has pursued a particularly promising technology for space hotels, in the form of inflatable habitats. This may sound a little risky if you think in earthbound terms, but remember that in space there really aren't many external stresses on a structure, so an inflatable one is perfectly adequate. It's also considerably lighter than a solid structure, which reduces the cost of launching it into orbit. On top of that, because it's flexible in its deflated form, it can be packed up into a relatively small space for transit. Once it's in its desired location, the walls can be inflated and the interior pressurised ready for its first visitors.

It's a technology that's already been tested in space with human occupants. In April 2016, a SpaceX Dragon cargo ship delivered a small demonstration unit – the Bigelow Expandable Activity Module, or BEAM – to the ISS. Readers in the UK may remember this particular Dragon visit, because it happened while British astronaut Tim Peake was on board the ISS. He used the station's remote manipulator arm to extract BEAM and attach it, rather like a prefab conservatory, to one of the station's existing modules. The following month, BEAM was successfully inflated to its full size, adding 16 cubic metres (or about 2 per cent) to the ISS's total pressurised volume.

The natural worry with an inflatable structure is that it will suddenly burst and let all the air out. But BEAM is designed to minimise this risk. Its skin is composed of a whole series of layers with different protective functions. On the outside are thermal insulation layers, then the air bladder that holds it rigidly in its inflated state. Then there are further layers designed to protect the module from micro-meteoroids and orbital debris. Even if all these layers fail, NASA claims that it wouldn't be a complete disaster. In their words, 'in the unlikely event of a puncture, BEAM would slowly leak instead of bursting. It is designed in this manner to preclude any damage to the rest of the space station.'

Bigelow has plans for a much larger 'production model', called B330 – and it's this that might end up serving as the world's first space hotel. It's essentially a scaled-up, standalone version of BEAM, with an internal habitable volume of 330 cubic metres. That's easy enough to remember, as it's where the module's name comes from (the B stands for Bigelow).

B330 won't require any complicated, ISS-style construction process, since the whole thing can be launched into space in one go, folded up to a third of its final size. As well as the pressurised living area, the unit includes solar panels, docking adapters and station-keeping thrusters. There's room inside for up to six people, though they'll have to put up with decidedly spartan living conditions – with little privacy, a limited space to practise zero-g acrobatics, and only tiny windows at best.

That's a far cry from the luxury space hotel that Arthur C. Clarke envisioned. So can we expect anything a little plusher than B330 in the near future? The honest answer is probably not, but that doesn't mean people haven't given any thought to the matter. There's no shortage of extravagant designs on paper, which are sadly never likely to see the light of day. Take the Von Braun Space Station, for example, 'announced' by an organisation called the Gateway Foundation in September 2019. It made it into the mainstream media at least as far as a British tabloid newspaper, *The Sun*, which is going to have to serve as our source on this.

The concept takes its name from Wernher von Braun, best known as the designer of both the V-2 and Saturn V rockets. In a magazine article in 1952, he described how a large, wheel-like space station could simulate gravity at its circumference if it produced enough centrifugal force via rotation. That's essentially how the space station described in *The Sun*'s article works. It would be 190 metres in diameter, with 24 comfortably furnished 'sleeping pods' in addition to restaurants, cocktail bars and a cinema. The Gateway

Foundation anticipates around a hundred visitors per week, and, according to *The Sun*, 'hopes to have its space station hotel up and running by 2025'. Good luck with that, then.

If a trip to an orbiting hotel doesn't sound exciting enough, what about a hotel on the Moon? It's an idea that goes all the way back to the dawn of the space age, with a Chicago newspaper claiming in August 1958 that 'the Hilton chain is dickering with the idea of opening the first hotel on the Moon.' In hindsight, no one seems to know if this notion really did originate with Hilton, or if it was just a product of someone's overactive imagination. In either case, by 1967 the famous hotel chain had decided it was a great publicity stunt. In that year, the *Wall Street Journal* reported that Barron Hilton had told them – quite possibly with tongue in cheek – 'that he was planning to cut the ribbon at an opening ceremony for a Lunar Hilton hotel within his lifetime'.

There was a time, during the 1960s and early 70s, when it was generally imagined that the next step after the Apollo landings would be to establish a permanent human presence on the Moon. So talk of a lunar hotel was, in some ways, less far-fetched then than it is today. The sad truth is that the big space agencies don't seem terribly keen on sending people back to the Moon at all (NASA has recently started hyping up a programme called Artemis, which ostensibly would do just that, but it's very much in the 'we'll believe it when we see it' category). The reason for NASA's attitude, and that of its counterparts in other countries around the world, is easy enough to understand. They only seem capable of thinking about space exploration in terms of scientific research, and

EXTRATERRESTRIAL INDUSTRIES

5

NASA has never liked the idea of space tourists. At the time of Dennis Tito's pioneering visit to the ISS in 2001, the agency's head, Daniel Goldin, was dead set against the idea. The practicalities of Tito's trip, in a Soyuz spacecraft, were arranged through the Russian space agency Roscosmos, but there was no escaping the fact that the 'I' in ISS stood for international. A large part of it was American-built, and there were NASA astronauts on board. When Tito and two Russian cosmonauts arrived at NASA headquarters for training on the US side of the station, they were sent packing. As a NASA official bluntly put it, 'We are not willing to train with Dennis Tito.'

But NASA isn't hostile to all forms of private enterprise. 'Serious' commercial activities, such as manufacturing – as opposed to frivolous ones like tourism – are a different matter. As long ago as 1973, NASA used its first space venture after the Moon landings, the Skylab orbital laboratory, to try out a number of potential fabrication processes such as growing crystals and mixing metals to produce alloys. They found that in cases like these, a 'zero-g' environment can offer significant advantages over Earth-based manufacture.

Of course, there are disadvantages to in-space manufacturing as well. You have to transport the raw materials up into orbit, and then return the finished products to Earth. But the consensus now, after several decades of research using Skylab, the Shuttle, Mir and the ISS, is that in several important instances the pros outweigh the cons. The aforementioned metal alloys represent one of these. Getting different metals to mix properly, especially if they include highly reactive elements like magnesium, is much easier in the absence of g-forces. We're not talking about mass-produced alloys for everyday applications, such as car wheels, but specialised ones for which there's a high demand, but in nothing like such large quantities. The super-strong, lightweight alloys used in medical implants are a good example.

Another product that may soon end up being made in space, rather than in earthbound factories, is fibre-optic cable. Everyone knows how central this has become to modern life, as one of the commonest ways to bring broadband internet into people's homes. It's a huge improvement over old-fashioned copper wire, but the fibre currently in use – essentially long strands of drawn-out glass – still suffers from significant transmission losses. In principle, the performance can be improved by making the fibre from a 'heavy-metal fluoride' glass, such as the dauntingly named

ZBLAN (because it contains fluorides of Zirconium, Barium, Lanthanum, Aluminium and Sodium).* But terrestrially fabricated ZBLAN ends up peppered with tiny crystals, which negate much of the performance gain. There's only way to make top-quality ZBLAN fibre, and that's to do it in the weightlessness of space.

Several commercial firms are hoping to get into the space manufacturing business, among them Axiom Space, Inc. You may remember that name from the previous chapter; it's the company planning to send Tom Cruise up in a Crew Dragon spacecraft to film a movie on the ISS. But Axiom sees 'microgravity research and manufacturing' as another key pillar of its future space business. In this context, it's looking at the viability of a number of options, including optical fibres, super-alloys and medical implants.

NASA's encouragement of 'extraterrestrial industry' doesn't stop with space manufacturing. It's also interested in the possibility of extracting and exploiting any natural resources that might be found beyond Earth. To this end, in September 2020, the agency issued a request for proposals from private companies for the supply of Moon rocks. These didn't have to be valuable minerals – any old rocks would do. And the companies didn't even have to bring them back to Earth, they could just stockpile them on the Moon for NASA to pick up at a later date. This may sound like a pretty pointless exercise, but from a symbolic point of view its significance was enormous.

---

* Sodium has the chemical symbol Na, which is why it's ZBLAN and not ZBLAS.

Exploiting the Moon

As with so many aspects of space travel, it all goes back to the Cold War. Remember the Outer Space Treaty, mentioned in the first chapter? Originating in 1967, this sought to place limits on 'the exploration and use of outer space, including the Moon and other celestial bodies', from the narrow perspective of the international politics of that time. Its authors couldn't conceive of a world that wasn't dominated by the nuclear arms race between the communist East and capitalist West, and all they really wanted to say was 'Please don't escalate that arms race into outer space'. But they wrapped it up in such legalistic overkill that what the treaty actually says goes much further than that.

The most contentious clause states that 'outer space is not subject to national appropriation by claim of sovereignty, by means of use or exploitation, or by any other means'. Everyone today understands that this really meant 'neither the Soviet Union nor the United States can claim ownership of the Moon, or an asteroid, in order to acquire a strategic advantage over the other side' – but, frustratingly, the actual words say much more than that. In one extreme reading, it could be interpreted to mean 'no individual or company can own any object originating in outer space' – the kiss of death for any future space prospecting industry.

Fortunately, few people are inclined to take such an extreme interpretation. The commonest view today is that the treaty puts outer space on the same footing as 'international waters' on Earth. No one can claim ownership of these waters, but anyone and everyone is free to use them. Anything they obtain there, such as seafood from fishing or valuable minerals from deep-sea mining, is theirs to keep or to sell on for a profit.

So that's where the NASA people were coming from with their 2020 call for proposals. They didn't really want the Moon rocks they were asking companies to collect for them. They simply wanted to validate the principle that, once collected, those rocks were the company's property, which they could legally sell on to NASA for whatever price was agreed.

In the event, NASA awarded four contracts to companies that submitted acceptable proposals, to a total value of just $25,000. That might sound like a fair price for a few rock samples, but don't forget you've got to design and build a spacecraft, pay to have it launched all the way to the Moon, and then have it land in one piece, collect the rocks, and cache them somewhere ready for NASA to pick up. You're not going to do that for a quarter-share in $25,000. They weren't equal shares, either. The bottom-rated of the four companies is just getting $1, and 80 per cent of that is only payable on delivery of the material. Never mind, it's the principle that counts – in this case the important principle that the company collecting the Moon rock has a legal right to sell it.

Overcoming the legal obstacles, however, is child's play compared to the technical challenge of getting a spacecraft safely down onto the lunar surface. To date, this has never been achieved by a commercial company, only by national governments – and even then, only three of the largest and richest. The United States, of course, famously landed humans on the Moon on six separate occasions between 1969 and 1972. That just happened to be the height of the Cold War – and around the same time, America's arch-rival, the Soviet Union, also managed to land a number of robotic probes.

These didn't make it into the history books in the way the US landings did, but they did achieve a couple of feats that are of comparable significance in retrospect. Three of the Soviet missions successfully returned rock samples to Earth, without requiring any human involvement at the lunar end, while another two deployed rovers to explore the terrain for much longer than a human mission could. In recent years, both these feats have been repeated by China – the third, and so far last, of the countries to have successfully landed on the Moon.

The difficulty in getting to the Moon isn't, as you might guess, because it's so far away. In fact, it doesn't require much more energy than putting a satellite into geostationary orbit, which is a common occurrence these days. The Chinese sample-return mission, Chang'e 5, had four separate spacecraft modules – an orbiter, lander, ascent vehicle and return capsule – totalling around six tonnes. Getting that to the Moon was actually quite easy, using a CZ-5 launcher of comparable performance to SpaceX's Falcon 9. The problem isn't with the size of the spacecraft, or the route it has to take – but with its sheer technical complexity. That's where all the money goes, not just in designing and building it, but in testing it and making sure everything functions exactly as it's supposed to. It's this complexity, and the associated expense, that has so far prevented commercial firms from following in the footsteps of the American, Soviet and Chinese governments.

In 2007, three years after SpaceShipOne won the X Prize that Peter Diamandis had set up for the first commercial suborbital flight, he announced a follow-up initiative. This time the subject was a robotic landing on the Moon, and the prize for the successful entry was doubled from $10 million to $20 million. For the original X Prize, the money had been provided by a wealthy family, the Ansaris, but for the Lunar X Prize it was the turn one of the world's richest companies, Google.

Up to a point, the rules were what you might expect. To win the prize, an entrant had to land a spacecraft on the Moon – paid for entirely through private investment – which would then send back pictures of its surroundings to prove it had done it. In an added twist, however, the rules stated that the spacecraft, or a surface rover deployed by it, then had to travel at least half a kilometre to a different location. This adds enormously to the complexity of the task. Historically, there was a gap of almost half a decade between the first soft-landing on the Moon, by the Soviet Union's Luna 9 in February 1966, and the deployment of the first robotic rover by Luna 17 in November 1970.

Nevertheless, plenty of companies took up the challenge, at least to the extent of designing spacecraft that could fulfil all the necessary requirements. Among the front-runners were Moon Express from the United States, SpaceIL from

Israel and Team Indus from India. Unfortunately, they all found that progress was slower than they'd hoped. As the original deadline of December 2014 approached with no launch in sight, the organisers offered a short extension, which was eventually stretched out to March 2018. Even that came and went without the prize being claimed, so reluctantly it had to be withdrawn.

One of the teams persevered even in the absence of a cash incentive. This was SpaceIL, with a lunar lander they called Beresheet, the Hebrew word for 'genesis'. Weighing just 585 kg, it could have been launched by quite a small rocket – but even so, paying for a dedicated space launch is still an expensive thing. So they took the alternative option of a 'ride share' into space on a Falcon 9, alongside another customer's much larger payload. This was a four-tonne geostationary satellite for an Indonesian telecommunications company, so we can assume that – if costs were allocated in proportion to mass – Beresheet got a pretty cheap ride into space.

The plan was to do everything the X Prize rules required, albeit to achieve that 500-metre relocation in a way the organisers probably hadn't envisaged. Instead of using a separate rover, the lander, once it was down, would briefly fire its engine again to hop to a new location. That's a trick that had been used once before, by NASA's Surveyor 6 lander in November 1967. It had a completely different reason for doing it, though: to demonstrate the feasibility of taking off from the lunar surface prior to the Apollo missions.

The question was, would Beresheet get a chance to repeat the trick, if only to meet the requirements of a prize that was no longer on offer? To start with, everything went according to plan. After that low-budget launch, on 22 February 2019, Beresheet was merely in Earth orbit, not on its way to the Moon. But it had a cunning plan to get there. Every now and then, the spacecraft fired its engine to put it into successively larger orbits, effectively spiralling outwards towards its destination. Eventually, on 4 April it arrived there, and fired its engine once again to enter lunar orbit.

A week later, Beresheet was ready for the most crucial step, the descent down to the surface. Unfortunately, this is

SpaceIL employees with a model of Beresheet, the first privately built spacecraft to reach the Moon, albeit in a crash-landing.
(Alon Hadar, CC-BY-SA-4.0)

where everything started to go wrong. A cascade of hardware and software failures meant that the lander's engine cut out long before it was supposed to, with the inevitable result. At 19.23 GMT on 11 April 2019, Beresheet became the first privately financed spacecraft on the Moon. But it was a crash, not a soft landing.

In spite of all the difficulties, there are still plenty of commercial incentives for going to the Moon. There's the prospect of lunar tourism, for one – whether in the form of that long heralded 'Lunar Hilton', or sightseeing tours of the historic Apollo landing sites, or the sporting possibilities opened up by a world which has just a sixth of Earth's gravity.

Another potentially lucrative option is the idea of lunar mining – extracting natural resources from the Moon. There are several reasons why people might choose to do this. One is for immediate local use, either by the aforementioned lunar tourism facilities, or by scientific outposts that may eventually be built on the Moon. Known by the generic term 'in-situ resource utilisation', or ISRU, the exploitation of locally sourced material is something that will become essential if humans are to establish more than a fleeting presence in outer space. We've already seen how difficult and expensive it is to launch mass into space from the Earth's surface. You really don't want to have to do that with all the fuel, food, oxygen and building materials that a lunar settlement is going to need.

At first sight, the Moon isn't a promising source of useful things like water and oxygen. In fact, it seems to consist of little else but grey igneous rock and dirt. The technical term for the latter is 'regolith', and it ranges from very fine dust to larger fragments of broken-up rock. You could think of it as lunar 'soil', although that term is generally limited to the Earth, where it has botanical connotations. Those certainly don't apply to the Moon, or to asteroids, where the term regolith is also used.

There's already quite a bit of lunar regolith, and larger Moon rocks, in laboratories here on Earth. The Apollo missions brought back over a third of a tonne of the stuff, supplemented in smaller amounts by the Russian and Chinese sample return missions. To everyone's surprise, some of the samples had quite a bit of water locked up in them – as much as five parts per million in some cases. That's the sort of thing that ISRU can work with. As well as the water itself, both the H_2O molecules and other minerals in the rock contain oxygen, and this is potentially extractable too.

In 2019, the European Space Agency (ESA) announced plans to send a robot lander to the Moon to test out a number of ISRU technologies. The goal of the mission is to demonstrate the feasibility of water and/or oxygen production on the Moon, with a stated target date of 2025. Closer to home, ESA has already set up a prototype ISRU plant at its research centre in the Netherlands, capable of extracting oxygen from lunar regolith.

If it turns out that commodities like oxygen and water can be obtained from Moon rock in large enough quantities, there's another possibility besides using them in situ – and that's transporting them back to Earth orbit for use there. This may sound unnecessarily complicated, when the same materials could be sourced just a few hundred kilometres away on the Earth's surface, instead of hauling them hundreds of thousands of kilometres from the Moon. But the difference in distance fades into insignificance when compared to the difference in the gravitational pull of the Earth and the Moon. Just compare the huge multi-stage rockets needed to reach orbit from the Earth with the much smaller spacecraft – such as the Apollo lunar module or the recent Chinese sample return mission – that can lift off from the Moon. As counterintuitive as it sounds, getting to Earth orbit requires far less effort if your starting point is the lunar surface.

Whether humans establish a thriving presence on the Moon in the near term is debatable. Far more certain is that we'll see an increasing amount of human activity in Earth orbit, both for the purposes of space tourism and industrial production. Providing all the necessary resources to support that activity will be easier if at least some of those resources can be sourced on the Moon rather than the Earth.

This was the thinking behind an ambitious but short-lived enterprise called the Shackleton Energy Company, founded in Texas in 2007. The aim was to set up processing plants on the Moon to extract water, and then electrolyse it into oxygen and hydrogen – the latter being one of the standard rocket fuels. The various products would then be transported to Earth orbit, where they would be sold to customers in need of them from the outer space equivalent of roadside service stations. It was a great idea, but one that was several decades ahead of its time. Aiming to finance itself through crowd-funding, the company only managed to raise a few thousand dollars before it was forced to call it a day.

Aside from extracting common elements like hydrogen and oxygen for use in space, there may be another, more glamorous, incentive for setting up a lunar mining operation. The lunar rock contains small traces of several elements, including yttrium and niobium, which are needed in high-tech applications like superconductors but are very rare on Earth. It could prove cost-effective to locate and extract them on the Moon and return them to Earth. In fact, one of the companies that went in for the Lunar X Prize, Moon Express, was originally set up with this particular end in mind.

There's another potentially valuable element, which can't be obtained for any amount of love or money here on Earth, but is there for the taking on the Moon – and that's helium-3 (see box). This may not be a topic of everyday conversation, but it's become a common trope in science fiction that depicts future industrialised lunar economies, from the 1992 novel *Assemblers of Infinity* by Kevin J. Anderson and Doug Beason to the brilliantly funny movie *Iron Sky* from 2012. As one of the characters says in the latter, helium-3 has the potential to make civilisation 'independent of all energy needs for the next thousand years'.

In the real world, a long-time advocate of helium-3 is Harrison Schmitt, the only scientist – as opposed to test pilot or member of the armed forces – ever to walk on the Moon.

A professional geologist, he was a member of the Apollo 17 crew that went there in 1972. As *Scientific American* reported in March 2017, 'for decades, he has championed the potential economics of lunar mining for helium-3, an isotope that could be crucial for certain forms of nuclear fusion.'

As the same article points out, the amount of helium-3 present in the lunar regolith is tiny – maybe 50 parts per million at best – which means that huge volumes of material would need to be processed in order to extract a usable quantity. Even so, with essentially zero helium-3 present on our own planet, if we're ever going to use it in fusion reactors then mining it on the Moon is still our best option.

What is Helium 3?

Many elements come in different forms, called isotopes, which differ at a nuclear level despite having identical chemical properties. Helium has two stable isotopes, helium-3 with two protons and one neutron in its nucleus, and helium-4 with a second neutron. The latter is the variety we find here on Earth, where atoms of helium-3 are virtually non-existent. They're created inside the Sun, and blown out into space with other particles in the solar wind. This never reaches the Earth's surface, however, because it's bounced back into space by our strong magnetic field. But the Moon doesn't have the same protective shield, and over the course of billions of years a comparatively hefty supply of helium-3 has built up in the lunar regolith.

This still doesn't explain why people make such a fuss over helium-3. The key lies in its industrial potential. In the world of energy supply, the brightest beacon on the horizon is nuclear fusion: potentially one of the cleanest and safest sources of energy, which might become a practical proposition in the decades to come. It turns out that helium-3 is the ideal fusion fuel, capable of producing large amounts of power with none of the harmful radioactivity usually associated with nuclear processes.

Asteroid Mining

The Moon isn't the only Solar System body that is relatively easy to reach. There are also thousands of 'near-Earth' asteroids – irregularly shaped chunks of rock that are mere tens or hundreds of metres across, or occasionally a few kilometres. In space, 'easy to reach' doesn't necessarily mean physically close, just that you can get there without consuming an enormous amount of rocket fuel. Near-Earth asteroids, for example, are ones that move on orbits around the Sun that are close to the Earth's orbit. On occasion these orbits may indeed bring them very near to the Earth, but at other times they can be 300 million kilometres away on the other side of the Sun. For a spacecraft, this distance is largely irrelevant. The fact that the *orbits* are close means it can get from Earth to the asteroid, and back again, with the expenditure of comparatively little energy. A human traveller might be put off by the fact that the resulting trip is likely to take many years, but robotic probes aren't bothered by such considerations. They can shut down most of their electronics and effectively cruise along in their sleep.

Another advantage of asteroids – over, say, the Moon – is their small size. For one thing, this means they produce almost no gravity, so 'landing' on them, and taking off again, requires very little energy. It also means you don't have to look very hard, or dig very deep, to find any valuable commodities they might possess. And it's these commodities that make asteroid mining such an attractive proposition. In terms of physical make-up, there are three broad types of asteroid: stony ones, carbonaceous ones and metallic ones. The last of these are the kind we're interested in.

The most desirable metals – historically in terms of social prestige, and now also for their hard-nosed industrial value – tend to be the heaviest ones. Soon after the Earth formed, when it was still in a semi-molten state, these all sank down to the centre of the planet. That meant the loss of all our native-born gold, platinum, palladium, osmium and the like. So why, you may ask, do we still have those things today, albeit in scarce supply? Answer: we got them from asteroids. These were much more plentiful in the early days of the Solar System, when they collided with the Earth quite often – replenishing, to some extent, the heavy metals that had been lost from its surface.

Now, however, we're running short of some of them, and that's inexorably pushing their prices up and up. Price, of course, is a function of two things: supply and demand. It just so happens that as the supply of heavy metals is going down, the demand for them is going up. People are finding more and more uses for them, in fields as diverse as micro-electronics, batteries, medical implants and catalytic converters for vehicles. If this demand continues to grow, and if the supply here on Earth continues to dwindle, then there will come a time when it really does become cost-effective to mine them from asteroids. There are plenty of heavy metals available there, as the following table shows.

Availability of some of the rarest elements on a typical metallic asteroid, together with the current selling price.

	Typical asteroid abundance (parts per million)	Price in dollars per gram (May 2021)
Iridium	16.0	202
Platinum	7.7	40
Palladium	3.7	95
Gold	2.8	59
Rhodium	2.1	90

Of course, if and when robots start to bring back tonnes of these elements from a convenient asteroid, they'd no longer be as scarce, and the price would start to drop. In a way, though, that's the whole point. It would mean the asteroid miners had cornered the market, and if they went out of business the price would shoot back up again. So it does look like a potentially profitable gamble for anyone brave enough to try it.

In a very small way, spacecraft have already been used to 'mine' asteroids – albeit only in the context of scientific

research by national space agencies, rather than as a commercial venture. The first attempt to collect asteroid material
was made by a Japanese probe, Hayabusa, which had a close
encounter with the near-Earth asteroid Itokawa – a stony
rather than metallic type – in 2005. Its attempt to scoop up
some regolith didn't go entirely according to plan, and when
the sample return canister was opened up on Earth it was
found to contain just a few grains of material.

More than a decade later, the follow-up mission
Hayabusa 2 was more successful. The target this time
was the carbonaceous asteroid Ryugu, and the spacecraft
collected not one but two samples. The first was regolith
scooped up from the surface, while the second was subsurface material that had been 'quarried' using a kind of gun.
The two samples, totalling around a hundred milligrams,
were successfully returned to Earth in December 2020.

Lagging somewhat behind Hayabusa 2, but with more
ambitious goals, is NASA's grandly named OSIRIS-REx –
for 'Origins, Spectral Interpretation, Resource Identification
and Security Regolith Explorer'. This was the spacecraft that
made headlines when it chomped up so much regolith – from
another carbonaceous asteroid, Bennu – that it couldn't close
the lid of the sampling head, causing it to regurgitate much
of its load back into space. Nevertheless, NASA believes it's
heading back to Earth with several hundred grams of material, far more than its Japanese predecessors.

As well as proving that it's possible to bring chunks of
an asteroid back to Earth, these missions have highlighted
another point: it's not something you can do in a hurry. In

each of the three cases, the total time from launch to arrival of the sample back on Earth added up to six or seven years. That's quite a deterrent to any private company hoping to make money from such a venture.

Like helium-3 extraction on the Moon, asteroid mining is something that is going to need more efficient technology than we have today before anyone can build a viable business around it. Hopefully that will happen within a decade or two, but for the present any plans in that direction are unlikely to lead anywhere. That's something that several companies have already discovered to their cost. Deep Space Industries and Planetary Resources, Inc., for example, both had great plans – and indeed great names – but the lead times and upfront funding were just too big to work out in a commercial environment.

One of the firms just named, Planetary Resources, was co-founded by a couple of people we've met already: Peter Diamandis, the brains behind the X Prize initiatives, and Eric Anderson, the CEO of Space Adventures. Like the Shackleton Energy Company, Planetary Resources hoped to set up a network of orbiting service stations – supplied, in this case, from raw materials obtained from asteroids instead of the Moon – to supplement the income from heavy metal mining. And also like the Shackleton Energy Company, they discovered pretty quickly that the business world wasn't ready for such ideas. 'Planetary Resources had to pause on its ambitions for mining asteroids and developing the resources of space because it's not a topic that is fundable yet,' as a company spokesman said in 2019.

Energy from Space

While we're talking about natural resources from space, it's worth mentioning the one thing in this category that life on Earth has been exploiting from the very start: solar energy. It's what warms the atmosphere to just the right temperature for us, and helps grow the plants that we use as food. In recent times, we've started using the Sun's energy in a more direct way, to generate electricity through solar panels. It's become one of the fastest-growing energy technologies in these carbon-conscious times, since it produces no harmful waste products at all. It's also fully sustainable, and effectively free – since the Sun produces all that energy anyway, regardless of whether we use it or not.

Even so, solar panels aren't as efficient as they might be. For one thing, the Sun is only in the sky during the hours of daylight, and even then (as British readers will attest) it can often be obscured by cloud. Even on a sunny day, only a fraction of the Sun's energy that impinges on the upper atmosphere actually reaches ground level. Much of it is absorbed on its way down, or reflected back into space. From the point of view of life on Earth, that's a good thing, because we don't want sunlight to be too intense, especially in the harmful ultraviolet wavelengths. But it does mean that solar panels are missing out on a lot of the energy that's otherwise there for the taking.

The fact is, solar panels work much more effectively in outer space. They're used as a matter of course by most satellites, space probes and the ISS to provide all their energy needs. So how about supplying energy to the Earth from space-based solar panels? It would have to be a pretty big array, hundreds of metres across to provide a gigawatt of power – enough to serve a large city – but that's not beyond the realms of possibility. The problems start when you have to get all that power down to the surface of the planet – presumably from a geostationary satellite, since it has to remain fixed relative to the rotating Earth.

The normal way to transmit electric power is through cables, but dropping a cable 36,000 km from a geostationary orbit isn't easy. We know, from the discussion of space elevators a couple of chapters back, that it can be done in theory – but in reality it's not going to happen for a long time. Fortunately, there's an alternative. The electricity can be converted to microwaves – the same kind of high-frequency radio waves used by mobile phones and microwave ovens – and these can then be beamed down to a receiving array on Earth. Because radio waves spread out as they travel, this array will need to be several kilometres across in order to catch all the energy. This means, for practical purposes, that it would have to be located offshore, or in a desert or similarly barren location. Nevertheless, it's an idea that is workable in principle, and one that many people have given serious consideration to.

Putting the solar panels in space has the obvious advantage of avoiding the 'cloudy day' problem, as well as atmospheric absorption and reflection. To a large extent it avoids the 'dark at night' problem too, because geostationary satellites are so high above the Earth's surface they're in direct sunlight most of the time. Even in the worst-case scenario – the middle of the night at the spring or autumn equinox, when the Sun is diametrically opposite the satellite on the other side of Earth – darkness lasts less than an hour and a quarter.

There's one potential problem with the system that may already have occurred to you. 'Hold on,' you may be thinking. 'An energy beam coming down from space, isn't that kind of like a death ray?' Well, it might be if something like a laser was used instead of microwaves, because a laser beam remains very tightly focused. Its energy would still be highly concentrated when it reached its destination, which could be pretty lethal to anything that got in its way. But a microwave beam doesn't behave like a laser; it spreads out over a larger and larger area the further it travels. By the

**A design concept for a 'solar concentrator' satellite
that would provide an almost continuous stream
of power at a receiver on the Earth's surface.**

(NASA image)

time it reaches the ground, the actual intensity per square metre, for any practical design, would be comparable to that of natural sunlight.

Several paper studies have concluded that space-based solar power, if it can be made to work, would be economically competitive with other power-generation technologies – particularly when they have to meet 'net zero' carbon targets. On the other hand, very little practical work has been done on the subject yet.

In 2019, the Chinese government announced a first step in this direction, with the construction of a 13-hectare facility to test the technique on a small scale. The experiment consists of an array of solar panels suspended from balloons at an altitude of a thousand metres, with the energy generated by the panels being beamed down to the ground in the form of microwaves. If this is successful, China hopes to have a full-scale version of the system – with the balloons replaced by geostationary satellites – operational by 2050.

This particular date is significant in the UK too, because it's when the government hopes to reduce the country's carbon emissions to net zero. Space-based solar power is on the Business Department's list of potential technologies to meet this goal, and the department has commissioned an engineering company, Frazer-Nash, to look into the option. As science minister Amanda Solloway said in 2020:

> This pioneering study will help shine a light on the possibilities for a space-based solar power system which, if successful, could play an important role in reducing our

emissions and meeting the UK's ambitious climate-change targets.

Although we're still several decades away from practical solar power satellites, the basic idea goes back a long way. It formed the basis for one of Isaac Asimov's earliest stories, 'Reason', first published in *Astounding Science Fiction* magazine in April 1941. The technical details differ slightly, but to all intents and purposes the principle that Asimov describes is the same one that people are considering today.

This isn't the first time in this book that we've encountered an idea that, while being perfectly feasible in theory and common currency in science fiction for years, has yet to see the light of day in the real world. Think about lunar colonies, or asteroid mining, or space elevators, or giant wheel-shaped space stations. The problem with all of these isn't that they're physically impossible, or that they could never become commercially viable. It's that they require an enormous amount of up-front investment – on things like research, development and testing – before they can reach the point where people start seeing profits from them.

This isn't a problem that science fiction writers have to worry about. Words are cheap, and they can constantly tantalise us with all the space-based technologies we ought to have by now, but don't. In the real world, however, the huge start-up costs involved in the space business are its biggest stumbling block. The question is, where are we going to find someone far-sighted enough – or naïve enough – to pick up the tab? That's what we're going to look at in the next chapter.

THE BILLIONAIRE SPACE RACE

6

As we've just seen, getting into the space business isn't cheap. With most new enterprises you can start small, and use the profits from initial sales to gradually expand your offering until you reach full-scale operation. With space, however, that's not an option. There's no such thing as a small or cheap rocket capable of reaching Earth orbit. Yes, there are relatively small and cheap rockets – but they're not going to reach space, and they're not going to earn you any money.

On top of that, there's all the research and development you need to do before you can even get off the ground. This means that, by the time you're in a position to offer a service to paying customers, the effective cost of those first offerings – and the price you have to charge the customer – is much higher than it will eventually be after all those up-front costs have been amortised. The same is true in any technology-based endeavour, but it goes double in space

– and you could end up pricing yourself out of the market before you've got going.

The upshot of all this is that, for all practical purposes, the space business is in the hands of a very rare type of individual. It has to be someone with a large amount of money – acquired through some other, safer enterprise than space – and the vision to put that money into an undertaking that might not see significant profits for several decades. They also have to be the kind of kids-at-heart who never grew out of the 'space is cool' phase. Welcome to the billionaire space race.

Of the various companies we've encountered in the course of this book, let's take another look at the most successful ones:

- SpaceX, the company behind one of the most successful rockets currently in operation, the Falcon 9 – used, among other things, to launch the company's own Starlink broadband satellites and to transport crews to the ISS in the Dragon capsule, also built by SpaceX. It's backed by billionaire Elon Musk, who made his initial fortune in internet banking, and now also runs the Tesla car company.
- Blue Origin, whose reusable New Shepard rocket is leading the race to provide affordable suborbital space tourism. The company was founded by Jeff Bezos, who made his billions from one of the most successful retail businesses in history, the internet giant Amazon.
- Virgin Galactic, also aiming at the suborbital tourism market with their innovative SpaceShipTwo spaceplane,

and Virgin Orbit, targeting the satellite launch sector with the air-deployed LauncherOne rocket. They're both parts of the sprawling Virgin empire founded by billionaire Richard Branson – which started out in the music business but has since expanded into the aviation, leisure, rail and telecommunications fields.

- Bigelow Aerospace, the only privately owned company to have contributed to the building of the ISS, with its inflatable BEAM extension module. The company also has plans to use the same technology to produce orbiting space hotels. It was in the earthbound hotel sector that its founder, billionaire Robert Bigelow, made his fortune.

The common factor, of course, is that word 'billionaire'. It's a different story when we look back at some of the companies we've encountered whose ambitious space projects never made it off the ground. There was Excalibur Almaz, for example – with its plans for tourist flights around the Moon – and the Shackleton Energy Company and Deep Space Industries, aimed at mining the Moon and asteroids respectively. Lacking big money of their own, these companies had to look elsewhere for investment – and, perhaps unsurprisingly, failed to find it.

The fact that space is such an expensive business has other ramifications too. Potential customers for space tourism, for example, are necessarily going to be very rich – and the rich have always been seen by certain unscrupulous types as potential targets for scamming. There's no evidence that any space tourism companies to date have been set up as deliberate scams, although it's an accusation that has been made in a couple of cases. On the other hand, as soon as reputable companies like Blue Origin and Virgin Galactic have established the market, it may be tempting for others to try to undercut them – either with an out-and-out scam, or with substandard technology. If things went really badly, something like that could destroy the market just as it was getting on its feet.

The fact that space tourists are inevitably going to be extremely wealthy has another downside, at least in the eyes of some people. When a European company, EADS Astrium, expressed a desire to enter the space tourism market in 2007, it was slapped down by the European Union's then industry commissioner, Günter Verheugen. As he explained to the Reuters news agency at the time, 'It's only for the super rich, which is against my social convictions.'

This is based on the false reasoning that a commercial business that targets the super-wealthy as its end customers will only be of financial benefit to such people. In fact, that's putting things back to front. The wealthy people are the ones that put money *into* the system, not the ones who walk off with the money. Space tourism ends up making its rich customers poorer, not better off. The people who really benefit financially aren't the company's customers, but its *employees*. The great majority of these will be in the kind of jobs to warm an EU commissioner's heart, such as factory workers and truck drivers.

The truth is that socialists don't have a monopoly on hating the super-rich. As a general rule, no one likes them very much. Fortunately, there are some exceptions to that rule, such as the new breed of space billionaires mentioned a moment ago. These are people who have a genuine concern for humanity's long-term future, and see space travel as an integral part of that future. They're not investing in it merely as a hard-nosed business proposition, but because they're following their dreams – and they're pretty big dreams, too. As X Prize founder Peter Diamandis succinctly put it, 'The meek shall inherit the Earth. The rest of us will go to Mars.'

Planet B

We've already seen, in the 'Vacations in Space' chapter, how Elon Musk hopes to send passengers to Mars and back using his giant Starship rocket. In fact, the Mars trip was one of the primary motivations for SpaceX developing Starship in the first place. At one point the company even referred to it as the 'Mars Colonial Transporter', and the choice of methane as a fuel was driven by the fact that stocks of it can be synthesised on Mars for the return trip.

Musk's obsession with Mars goes back a long way. In 2001 – the year the first space tourist, Dennis Tito, flew to the ISS – SpaceX hadn't yet been created, and Musk was best known as an internet entrepreneur. But even then, he revealed his ambition to set up the first human settlement on Mars, describing it as 'a positive, constructive, inspirational goal capable of uniting humanity at a critical time'.

It took Musk fifteen years to transform this generalised aspiration into a specific plan for getting to the Red Planet. He revealed this to the world at the 67th International Astronautical Congress (IAC), held in Mexico in September 2016. In all essential details, the plan described by Musk on that occasion was identical to the current one using Starship – except for the name, which at that time was the rather clunky 'Interplanetary Transport System'.*

At the 67th IAC, Musk also explained why, in his view, getting humans to Mars is such a high priority endeavour. For him, the Red Planet isn't simply an exciting new tourist destination – or, as it is for NASA, a scientific curiosity. It's humanity's Planet B, and over the course of the coming centuries it may prove essential to the survival of our species. In his words:

> I think there are really two fundamental paths. History is going to bifurcate along two directions. One path is we stay on Earth forever, and then there will be some eventual extinction event. I don't have an immediate doomsday prophecy – just that there will be some doomsday event. The alternative is to become a space-faring civilisation and a multi-planet species.

A year later, in September 2017, the 68th IAC was held in Australia – and once again Musk talked about SpaceX's Mars plans. More specifically, he talked about the elephant in the

---

* For further details, see *Destination Mars*, also in the Hot Science series from Icon Books.

room that he'd ignored the previous year. As he put it on this occasion: 'I think we've figured out how to pay for it.'

The key to Starship's profitability is the fact that it's fully reusable, not just partially so like Falcon 9 and Falcon Heavy. This means that, once developed, it will be a more cost-effective way to carry out SpaceX's core work, like launching satellites or ISS crews. In other words, Starship has more strings to its bow than long-haul flights beyond Earth orbit. It was precisely the lack of such flexibility that killed off the only other rocket that's ever been in Starship's class, NASA's Saturn V.

Musk's idea is that the profits from Starship's 'everyday' work will cover all its development costs, as well as going part way towards financing the (potentially much less profitable) Mars missions. He even has his eye on a completely new market for the giant rockets, one that doesn't even involve going into Earth orbit. This is point-to-point travel from one geographic location to another, using a suborbital trajectory. 'Most of what people consider to be long-distance trips could be completed in less than half-an-hour,' Musk told his audience in Australia.

This practical approach, thinking through not just the technicalities of getting to Mars but the financial aspects too, is what sets Elon Musk apart from countless dreamers before him. As Alan Duffy, a professor from Melbourne, said after the IAC talk:

What I love about SpaceX, and why the world's scientists and engineers are willing to give them credence, is that they make things profitable at every step of the way. They have a big vision, they work towards it, but the steps they take are always with profit in mind. And if there's a profit there, you can guarantee that businesses will see it through.

Of course, there's still a lot of work to be done before the first passenger-carrying Starship goes to Mars. The rocket – which you'll remember is in two stages, the Starship proper and the 'Super Heavy' booster – has to be fully tested and certified for human flight. That's likely to take several years. And you can't just go to Mars any time you like. If you've read *Destination Mars* in this series, you'll know that there is just one brief 'launch window', when Mars and Earth line up in just the right way, every 26 months or so. Even then, the journey between the two planets is going to take around six months, and several robotic missions – to set up the necessary resources at the Mars end – will be needed before it's safe for humans to make the trip.

The first Starships to land on Mars won't be packed with people. The maximum seating capacity is a hundred – enormous in comparison to any previous spacecraft – but that would only be used on 'short-haul' flights, such as those hops from one side of Earth to the other, or orbital excursions lasting just a few days. To give people enough space to live comfortably on the long trip to Mars, the passenger complement would be reduced to a few dozen – or even further than that, to ten or twelve, on the earliest flights, because so much space will be needed for cargo.

In effect, the first tourists on Mars will have to take their hotel along with them, in kit form, and put it together for themselves.

Actually, 'tourist' may not be quite the right word here. It's true that some of them will be fare-paying passengers who go for a few weeks and then return to Earth, but others will be there to do a long-term job. After all, someone is going to have to build a settlement on Mars that is more than just a temporary shelter from the elements.

An artist's conception of an early settlement on Mars.
(NASA image)

These first Martian settlers won't be SpaceX employees – they'll be working for themselves, or for some other private company. Musk sees SpaceX purely as a transportation company, ferrying people – ultimately, he hopes, for as little as $200,000 per person – and cargo between Earth and Mars. His aim is to make it as easy as possible for other players to create a permanent base on the Red Planet.

That's not to say Musk doesn't have ideas of his own on how such a base might be constructed. In a Reddit 'Ask Me Anything' session in October 2016, he described how a mixture of surface and subsurface structures could be constructed using materials sourced locally, as well as those transported from Earth:

> Initially, glass panes with carbon-fibre frames to build geodesic domes on the surface, plus a lot of tunnelling droids. With the latter, you can build out a huge amount of pressurised space for industrial operations and leave the glass domes for green living space.

So that's Elon Musk's vision of how our 'Planet B' might be created. Other people, however, are thinking along different lines.

The High Frontier

If Elon Musk has an arch-rival in the space business, then it has to be the founder of Blue Origin, Jeff Bezos. The two men have a lot in common. They were both born around the time of the first space race – Bezos in 1964, Musk in 1971 – and both cut their business teeth in the 'dot com' boom of the 1990s. Like Musk, Bezos can trace his interest in space back a long way. In 1982, when he was still a student at a Florida high school, a local newspaper described his ambition 'to build space hotels, amusement parks and colonies for two or three million people who would be in orbit'.

Reading that again slowly, we can see two completely different ideas in there. 'Hotels and amusement parks' sounds not too different from the sort of space tourism that Blue Origin is currently getting into. But 'colonies for two or three million people'? That's several orders of magnitude more grandiose, and it's really another variation on the 'Planet B' theme. Elon Musk's postulated 'extinction event' aside, that's something that we're going to need eventually anyway, if the Earth's population of humans continues to grow and grow. As Bezos said in that 1982 interview, 'The whole idea is to preserve the Earth.'

After graduating from that Florida high school, Bezos went to Princeton University to study for a Bachelor of Science in Engineering degree. By an amazing coincidence, one of his professors turned out to be the world's leading authority – which in those days meant 'only authority' – on large orbiting space habitats. This was Gerard O'Neill, who in 1976 had produced a book called *The High Frontier: Human Colonies in Space*. Like the teenage Bezos, O'Neill's motivation was a mixture of enthusiasm for space and concern for the terrestrial environment in the face of an ever-increasing human population.

O'Neill's favoured approach was a cylindrical habitat, now referred to as an O'Neill cylinder in his honour. Like the wheel-shaped space stations of science fiction, this would use the centrifugal force of rotation to produce an artificial gravity effect on the inside surface. Unlike any space station, however, the cylinders of O'Neill's imagination were gigantic. His basic design called for a diameter of 8 km and a length of 32 km – which, if you work it out, gives a total internal surface area of just over 800 square kilometres. That's slightly larger than the island of Singapore, which has a population of 5.7 million.

O'Neill went into considerable detail on the technicalities of his space habitats. For example, they had to be close enough to the Earth to be easily accessible, but far enough away that there was no risk of them ever crashing down to the surface of the planet. On the other hand, the material

An imaginative view of what a large space habitat, such as an O'Neill cylinder, might look like from the inside.
(NASA image)

to construct them had to come from somewhere that didn't require huge amounts of energy to launch mass into space – and that sounds more like the Moon than the Earth. O'Neill concluded that the best location was a spot in the same orbit as the Moon, but either 60 degrees ahead of it or behind it. Technically referred to as Lagrange points 4 and 5, these are known for their long-term stability as well as being easily accessible from both the Earth and the Moon.

When Gerard O'Neill's book appeared in 1976, 'space travel' in the Western world effectively meant 'NASA or nothing'. As it happens, NASA did take a brief interest in his ideas. After the Moon landings, the agency dithered for several years trying to work out what it wanted to do next, and an O'Neill-style space habitat was one of the options under consideration. However, apart from a series of fanciful artist's impressions, nothing ever came of it.

Fast forward to May 2019, when Jeff Bezos revealed his own counterpart to Elon Musk's Mars vision, at a debate on space colonisation in Washington DC. Not surprisingly, this involved O'Neill cylinders – lots and lots of them, perhaps enough eventually to house a trillion inhabitants. Obviously, it would take centuries to build them all, presumably by cannibalising virtually the entire Moon, but the idea does have some kind of logic on its side. The fact that 'gravity' is generated by the centrifugal effect of rotation, rather than in the usual way by the attraction of a massive body, means the quantity of matter needed to support a given number of people is far less than it would be on the surface of a planet. In other words, the same amount of material goes much further when it comes to creating usable living space.

A purpose-built habitat has other advantages too. The occupants would have complete control over the weather, and natural disasters would be virtually non-existent (the cylinder could be equipped with thrusters to get it out of the path of an incoming asteroid, for example). And the cylinders don't have to be identical; designers can use their imagination. Some might have very Earth-like landscapes and architecture, while others could be worlds of fantasy. To quote Bezos:

> Some of them would be more recreational. They don't all have to have the same gravity; they can have a recreational one that keeps it zero-g so you can go flying with your own wings.

The idea of transporting material from the Moon to build free-floating colonies in space makes a lot of sense in the long term, where 'long' means several centuries. Viewed on a shorter timescale, however, it borders on the downright silly. Why not just build the habitat on the lunar surface and dispense with all the transportation problems? One of the first people to point this out after Bezos's speech was none other than Elon Musk, who tweeted: 'Makes no sense ... would be like trying to build the USA in the middle of the Atlantic Ocean.'

While Musk and Bezos both want humanity to expand outwards into space, their favoured routes are completely different. It's a character conflict worthy of a novel. To top it off, they're not just two starry-eyed dreamers. At the time of writing, according to the Bloomberg Billionaires Index,* they are the two richest individuals on Earth. It's worth taking a moment to let that sink in. The two wealthiest people in the world are both space-mad geeks who see humanity's future lying somewhere other than its home planet. If someone like Arthur C. Clarke had put that in a sci-fi novel 50 years ago, everyone would have said it was too far-fetched to be credible.

What's not far-fetched at all, given human nature, is the fact that Musk and Bezos can't agree on the way forward. If they did, and if they pooled their fortunes and went all out for the same goal, maybe we'd get there that much more quickly. On the other hand, this is a book about private enterprise, and that's something that has always thrived on competition. So perhaps things are better this way after all.

When people talk about the billionaire space race, they're usually thinking of Jeff Bezos, Elon Musk and a third contender – Richard Branson, with his Virgin Group of companies. The latter includes both Virgin Galactic, which is in direct competition with Bezos's Blue Origin in the suborbital tourism market, and Virgin Orbit, which aims to take on SpaceX and others in the satellite launching business. In some ways, however, Branson is the odd one out of the three, being UK- rather than US-based, and having started his business career in pre-internet times. Virgin's first ever

---

* https://www.bloomberg.com/billionaires/

product, the iconic Mike Oldfield album *Tubular Bells*, was released in 1973, when Elon Musk was just two years old.

Branson's long-term space ambitions are different from those of Musk and Bezos, too. He rarely talks about the Moon or Mars or orbital space habitats. In fact, he admits that Virgin is 'more interested in how we can use space to benefit the Earth'. This is driven by the same environmental worries that are prompting Bezos and Musk to seek out a 'Planet B', but with a completely different approach. Branson believes that, by making suborbital flights as affordable as possible, more and more people will see for themselves just how 'beautiful and in need of protection' our planet is.

Sailing to the Stars

This 'cosmic perspective' is a by-product of the scientific age. In earlier centuries, people didn't view the Earth as a planet at all, but as the fixed centre of the universe. It's only with the invention of telescopes that we've learned that all the stars in the sky are distant suns that might have their own planets orbiting around them. But as much as we can appreciate that intellectually, we're still stuck with this one viewpoint from inside the Solar System. The other stars are all so far away, we could never hope to get a close-up view of even one of them. Or could we?

The surprising answer is yes we can, and it might even happen within the lifetime of people reading this book. It's time to meet yet another billionaire – a Russian this time, named Yuri Milner. In 2015 he set up a series of 'Breakthrough' initiatives, aimed at expanding humanity's cosmic horizons in various ways. The initiative we're concerned with here is called Breakthrough Starshot, and it involves sending a huge fleet of spacecraft to the Alpha Centauri system four light years away. While the venture may be lacking in any obvious commercial potential, the intention is to fund it through private investment, so it legitimately falls within the remit of this book. In any case, it's such an audacious concept it's worth looking at it in some detail.

The Starshot craft – several hundred of them – would be travelling at around 200 million kph, or 20 per cent of the speed of light, which means it would take them just 20 years to travel those four light years. That compares very favourably to the 70,000 years that NASA's Voyager probes will take to cover the same distance. After reaching their destination, it will take another four years for the Starshot data, travelling at the speed of light, to get back to Earth. Adding a lead time of fifteen years to set the whole thing up, that's a total elapsed time – starting at the present moment – of just 39 years.

It sounds like pure science fiction, but there are a few sneaky tricks, made possible by modern technology, that turn it into a viable proposition in the real world. For one thing, the spacecraft will be tiny – even smaller than cube-sats. Needless to say, there will be no humans on board, just miniaturised sensors and microprocessors. And the probes aren't going to stop, or even slow down, when they reach their destination – simply whiz past at that incredible rate of 200 million kph. In one final twist, they're not going to achieve this eyewatering speed using rocket engines, but sails.

Before looking at Milner's proposal in any more detail, it's worth taking a brief detour into the strange-sounding but well-established physics underlying Breakthrough Starshot. You'll remember that a rocket works by exploiting Newton's third law of motion. It builds up forward momentum by expelling an exhaust stream with equal and opposite momentum in the reverse direction. That's the only way a self-contained spacecraft can accelerate to high velocity.

On the other hand, if it's not a self-contained system, there's another way a craft can gain momentum. Think of the analogy of a sailing ship. It doesn't have any motive power of its own, but it's pushed along by the wind. In physics terms, the air molecules transfer some of their momentum to the ship's sails. Might it be possible to put the same principle to work in space? All we need is a suitable external source of momentum.

As we saw when we talked about solar power satellites, the Sun's light contains a huge amount of 'free' energy that we can exploit using solar panels. It turns out that this light carries momentum, too. Technically called 'radiation pressure', it's too weak to be discernible in an everyday context. Even on a very sunny day, when you can easily feel the Sun's energy on your face, you can't feel its radiation pressure. But it's there, and objects out in the vacuum of space can feel it. When NASA sent the first Viking landers to Mars, for example, they had to compensate for the effects of radiation pressure. If they hadn't, the craft would have been blown off course, missing the Red Planet by 15,000 km.

But could solar radiation pressure ever be used as a spacecraft's primary means of propulsion? To be practical, the spacecraft would have to be very small and light, but in these days of cubesats that's much less of a constraint than it once was. In fact, a proof-of-concept demonstrator, called LightSail 2, has already been flown successfully. It was designed and built by the Planetary Society, an international non-profit organisation with a vision to 'know the cosmos and our place within it'.

Launched along with two dozen other satellites by a SpaceX Falcon Heavy in June 2019, LightSail 2 is a three-unit cubesat – in other words, a small rectangular box with dimensions $10 \times 10 \times 30$ cm. After settling into orbit at an altitude of 720 km, the satellite unfurled its huge sail, a thin Mylar square 5.6 metres on a side. It was then able to use this, rather than the thrusters employed by conventional satellites, to move from one orbit to another – all thanks to the radiation pressure of the Sun.

As useful as 'solar sailing' is to a cubesat manoeuvring in Earth orbit, it's never going to get it to the stars. The momentum boost obtained from solar radiation is tiny. In the case of LightSail 2, if it could be kept up for a month, all in the same direction, it would only increase its speed by 550 kph. If we want to use radiation pressure to get anywhere near the speed of light, we're going to need a much more intense source of light than the Sun.

**Artist's rendering of the Planetary Society's
LightSail 2 after deployment in Earth orbit.**
(The Planetary Society, CC-BY-SA-3.0)

This brings us back to Yuri Milner and the Breakthrough Starshot project. His solution is to use an array of ground-based lasers with a combined power of a hundred gigawatts.* This would be aimed at Alpha Centauri, and used to accelerate – one by one – each of the spacecraft in the interstellar 'fleet'. These would have sails similar in size to LightSail 2, but the spacecraft themselves would be even tinier, weighing just a few grams each. After working through the maths, it turns out that the laser beam could accelerate them to the desired velocity – a fifth of the speed of light – in just ten minutes of continuous operation.

---

* A gigawatt is a million kilowatts.

Of course, aiming a laser at a star system four light years away isn't going to be that precise – hence the scattershot approach of using hundreds of tiny spacecraft. The chances are that at least a few of them will get close enough to snap some interesting photographs. Thanks to today's miniaturised technology, they would have all the necessary equipment on board to do that, and send the pictures back to Earth via their own tiny laser transmitters.

The implications are jaw-dropping. In contrast to countless previous speculations about interstellar travel, the numbers here – minutes, grams, even that hundred-gigawatt laser beam – actually look doable within a realistic space-project budget. Milner estimates the total cost at between five and ten billion dollars to meet a launch date of 2036.

Of course, the project doesn't have the potential customer base of, say, space tourism. In fact, in the traditional sense, it doesn't have a customer base at all. But the sheer symbolism of a successful trip to the stars, with the uplifting effect it would have on the whole of humanity, is something that a certain type of philanthropist would be happy to dig into their pockets for. Facebook billionaire Mark Zuckerberg, for one, has been enthusiastic about the Starshot project since its inception.

All things considered, the billionaire space race has some pretty ambitious goals. From Martian settlements to O'Neill cylinders to fleets of miniature spacecraft crossing the four light years of space to Alpha Centauri, you can't accuse its key players of thinking small. And you never know, one or more of those dreams might actually come to pass in the decades to come. On the other hand, they might all disappear without a trace.

The problem with dreams is that they tend to be personal. If the individual with the vision – whether that's Elon Musk, Jeff Bezos or Yuri Milner – can maintain the necessary finances and support to see it through to completion, then yes, it might happen. If not, then the likelihood is that subsequent generations will have their own dreams. Maybe these too will involve expanding outwards from Earth, but they may not. They may follow Richard Branson's example instead, seeing the priority as preserving and improving our own planet, rather than seeking out a 'Planet B'. The increasing power of the environmental movement suggests that perhaps this is a more plausible future.

Long-term visions aside, the fact is that the space business is here to stay. It's simply too useful, even from a completely Earth-centric perspective, for the modern world to dispense with it now. Though we don't necessarily realise it on a day-to-day basis, we just couldn't live the way we do if we didn't have satellites – for everything from navigation and communication to weather forecasting and climate monitoring. And, with the advent of broadband constellations like SpaceX's Starlink, it's a trend that is growing exponentially.

Then there's space tourism. As it becomes better established and more affordable, an increasing number of people are going to take vacations that way, even if it's just a few days in an orbiting hotel to see the Earth from a different point of view. At the same time, with more and more countries striving for 'net zero' carbon emissions, things like solar power satellites – and perhaps even nuclear fusion using helium-3 mined on the Moon – are going to look like increasingly attractive possibilities. So if you're an entrepreneur or investor in search of a sure-fire bet for the future, look no further than the space business.

FURTHER READING

General Interest

Brian Clegg, *Final Frontier: The Pioneering Science and Technology of Exploring the Universe* (St Martin's Press, 2014)

Joseph N. Pelton, *The New Gold Rush: The Riches of Space Beckon!* (Springer, 2017)

Space Tourism

Blue Origin, 'Become an Astronaut': https://www.blueorigin.com/new-shepard/become-an-astronaut/

Anthony Cuthbertson, 'SpaceX Reveals Civilian Passengers for Trip into Space this Year', *The Independent*, 2 April 2021: https://www.independent.co.uk/life-style/gadgets-and-tech/space/spacex-launch-2021-space-dragon-b1825974.html

Stephen Clark, 'Space Adventures Announces Plans to Fly Private Citizens on SpaceX Crew Capsule, *Spaceflight Now*, 18 February 2020: https://spaceflightnow.com/2020/02/18/space-adventures-announces-plans-to-launch-private-citizens-on-spacex-crew-capsule/

Mike Fleming, 'Out Of This World: Tom Cruise Plots Movie to Shoot in Space with Elon Musk's SpaceX', *Deadline*, 4 May 2020: https://deadline.com/2020/05/tom-cruise-movie-shot-in-outer-space-elon-musk-spacex-unprecedented-in-hollywood-1202925849/

Paul Rincon, 'What is Elon Musk's Starship?', *BBC News*, 4 March 2021: https://www.bbc.co.uk/news/science-environment-55564448

Tariq Malik, 'How SpaceX's First Passenger Flight Around the Moon with Yusaku Maezawa Will Work', *Space.com*, 18 September 2018: https://www.space.com/41856-how-spacex-bfr-moon-passenger-flight-works.html

Jonathan Owen, 'Shooting for the Moon: Time Called on Isle of Man Space Race', *Independent*, 11 March 2015: https://www.independent.co.uk/news/science/shooting-moon-time-called-isle-man-space-race-10101750.html

Megan Barber, 'You Could Stay in a Space Hotel Pod by 2021', *Curbed*, 23 February 2018: https://archive.curbed.com/2018/2/23/17043960/space-hotel-bigelow-aerospace

Charlotte Edwards, '"World's First Space Hotel" Revealed with Artificial Gravity and Stunning "Earth View" Cabins for 400 Astro-tourists', *The Sun*, 2 September 2019: https://www.thesun.co.uk/tech/9846549/worlds-first-space-hotel-revealed/

SpaceX, 'Mars and Beyond': https://www.spacex.com/human-spaceflight/mars/

Jonathan Amos, 'Elon Musk: Rockets Will Fly People from City to City in Minutes', *BBC News*, 29 September 2017: https://www.bbc.co.uk/news/science-environment-41441877

Brian McGleenon, 'The O'Neill Cylinder: Jeff Bezos' Vision for an Incredible Civilisation in Space Supporting Entire Ecosystems', *Medium*, 27 August 2019: https://medium.com/@lynwerkledges/the-oneill-cylinder-jeff-bezos-vision-for-an-incredible-civilisation-in-space-fef75b499710

Satellites and Launchers

Iridium Communications Inc., 'Satellites 101: LEO vs. GEO': https://www.iridium.com/blog/2018/09/11/satellites-101-leo-vs-geo/

Darrell Etherington, 'SpaceX Hopes Satellite Internet Business Will Pad Thin Rocket Launch Margins', *TechCrunch*, 13 January 2017: https://techcrunch.com/2017/01/13/spacex-hopes-satellite-internet-business-will-pad-thin-rocket-launch-margins/

Shannon Hall, 'After SpaceX Starlink Launch, a Fear of Satellites that Outnumber all Visible Stars', *New York Times*, 1 June 2019: https://www.nytimes.com/2019/06/01/science/starlink-spacex-astronomers.html

John Maguire, 'Tiny Satellite Aims to Inspire Schoolchildren', *BBC News*, 25 November 2013: https://www.bbc.co.uk/news/av/science-environment-25084547

Loren Grush, 'A 3D-Printed, Battery-Powered Rocket Engine', *Popular Science*, 14 April 2015: https://www.popsci.com/rocket-labs-got-3d-printed-battery-powered-rocket-engine/

Chris Gebhardt, 'NASA's ICON Mission Launches on Northrop Grumman Pegasus XL Rocket', *NASA Spaceflight*, 10 October 2019: https://www.nasaspaceflight.com/2019/10/nasas-icon-launch-ngis-pegasus-xl-rocket/

Natalie Clarkson, 'Virgin Orbit One Step Closer to Launches from Spaceport Cornwall', Virgin.com, 4 May 2021: https://www.virgin.com/about-virgin/latest/virgin-orbit-one-step-closer-to-launches-from-spaceport-cornwall

Reaction Engines, 'SABRE: The Engine that Changes Everything': https://www.reactionengines.co.uk/beyond-possible/sabre

Jason Daley, 'Japan Takes Tiny First Step Toward Space Elevator', *Smithsonian Magazine*, 5 September 2018: https://www.smithsonianmag.com/smart-news/researchers-take-tiny-first-step-toward-space-elevator-180970212/

The Planetary Society, 'What Is Solar Sailing?': https://www.planetary.org/articles/what-is-solar-sailing

Maddie Stone, 'Stephen Hawking and a Russian Billionaire Want to Build an Interstellar Starship', *Gizmodo*, 4 December 2016: https://gizmodo.com/a-russian-billionaire-and-stephen-hawking-want-to-build-1770467186

Industrialisation of Space

Prachi Patel, 'Four Products that Make Sense to Manufacture in Orbit', *IEEE Spectrum*, 26 November 2019: https://spectrum.ieee.org/aerospace/space-flight/4-products-that-make-sense-to-manufacture-in-orbit

Elizabeth Gibney, 'Israeli Spacecraft Beresheet Crashes into the Moon', *Nature*, 11 April 2019: https://www.nature.com/articles/d41586-019-01199-2

Katia Moskvitch, 'Much More Water Found in Lunar Rocks', *BBC News*, 14 June 2010: https://www.bbc.co.uk/news/10313173

Jonathan Amos, 'China's Chang'e-5 Mission Returns Moon Samples', *BBC News*, 16 December 2020: https://www.bbc.co.uk/news/science-environment-55323176

Leonard David, 'Red Planet versus Dead Planet: Scientists Debate Next Destination for Astronauts in Space', *Scientific American*, 30 March 2017: https://www.scientificamerican.com/article/red-planet-versus-dead-planet-scientists-debate-next-destination-for-astronauts-in-space/

Paul Rincon, 'Hayabusa-2: Capsule with Asteroid Samples in Perfect Shape', *BBC News*, 6 December 2020: https://www.bbc.co.uk/news/science-environment-55201662

Chloe Cornish, 'Interplanetary Players: a Who's Who of Space Mining', *Financial Times*, 19 October 2017: https://www.ft.com/content/fb420788-72d1-11e7-93ff-99f383b09ff9

Alan Boyle, 'One Year after Planetary Resources Faded into History, Space Mining Retains Its Appeal', *GeekWire*, 4 November 2019: https://www.geekwire.com/2019/one-year-planetary-resources-faded-history-space-mining-retains-appeal/

Tom Bawden, 'The Final Frontier for Energy', *The I*, 14 November 2020 (print only – transcript available at https://beastrabban.wordpress.com/2020/11/17/i-british-government-considering-solar-power-satellites/)

Scott Snowden, 'China Plans to Build the World's First Solar Power Station in Space', *Forbes*, 5 March 2019: https://www.forbes.com/sites/scottsnowden/2019/03/05/china-plans-to-build-the-worlds-first-solar-power-station-in-space

INDEX